BHB

Advances in Surface Engineering
Volume I: Fundamentals of Coatings

Advances in Surface Engineering
Volume I: Fundamentals of Coatings

Edited by

P.K. Datta
University of Northumbria at Newcastle, UK

J.S. Burnell-Gray
University of Northumbria at Newcastle, UK

THE ROYAL
SOCIETY OF
CHEMISTRY
Information
Services

The Proceedings of the Fourth International Conference on Advances in Surface Engineering held at The University of Northumbria at Newcastle on 14–17 May 1996.

The front cover illustration is taken from the contribution by J. Kipkemoi and D. Tsipas, p.32.

Special Publication No. 206

ISBN 0-85404-747-6

A catalogue record for this book is available from the British Library

Published by The Royal Society of Chemistry,
Thomas Graham House, Science Park, Milton Road,
Cambridge CB4 4WF, UK

Printed and bound by
Bookcraft (Bath) Ltd

Preface

Advances in Surface Engineering is based on the Proceedings of the *4th International Conference on Advances in Surface Engineering* which was hosted by the University of Northumbria's *Surface Engineering Research Group* between 14–17th May 1996.

Acknowledgements

The editors wish to thank Professor Gilbert Smith, the Vice Chancellor of the University of Northumbria for opening the conference. The editors are grateful to Professor Van de Voorde for giving the opening technical keynote talk.

The editors wish to express their gratitude for the support extended by:

The Department of Trade and Industry, The Institute of Materials, The Institute of Corrosion, The Royal Society of Chemistry, Multi-Arc (UK) Ltd, Buehler Ltd, Gearing/Micromaterials, Mats (UK), Tech Vac and Woodhead Publishing.

Special thanks are due to Professor Cryan head of the *School of Engineering* at the *University of Northumbria at Newcastle*.

The support and encouragement of many colleagues at the *University of Northumbria at Newcastle* and friends in other universities, is gratefully acknowledged.

The human commitment to any conference or book is substantial and often not fully acknowledged. In this regard the work of Kath Hynes, the secretaries and the technicians from the *School of Engineering* and the members of the *Surface Engineering Research Group* (*SERG*) should be fully recognized.

Finally special commendation is reserved for Dan Smith who administered the *4th International Conference on Advances in Surface Engineering* (*4ICASE*) and David Griffin of *SERG* who desk-top published the conference proceedings.

P. K. Datta
J. S. Burnell-Gray

Surface Engineering Research Group
School of Engineering
University of Northumbria at Newcastle

Contents
Volume I Fundamentals of Coatings

Section 1.3 Wear

Section 1.4 Fatigue and Other Failure

Contents
Volume II Process Technology

Introduction

Section 2.1 PVD and CVD

Section 2.2 Thermal, Plasma, Weld and Detonation

Section 2.5 Electrochemical and Electroless

Contents
Volume III Engineering Applications

Introduction

Section 3.1 Biomedical

Section 3.2 Aerospace

Introduction

J. S. Burnell-Gray and P. K. Datta

SURFACE ENGINEERING RESEARCH GROUP, UNIVERSITY OF NORTHUMBRIA AT
NEWCASTLE, UK

1 SCOPE OF *ADVANCES IN SURFACE ENGINEERING*

Advances in Surface Engineering is aimed at reviewing and documenting the recent advances in research and application of this relatively newly emerging technology, the problems that remain to be solved and the directions of future research and development in surface engineering.

The three volumes incorporate both science and technical research papers. They demonstrate how SE technologies are continuously increasing the level of performance of components, devices and structures through the creation of high performance surfaces.

What distinguishes *Advances in Surface Engineering* is an attempt, though limited, to characterise the coatings/engineered surfaces in terms of their fundamental structural entities and to understand their behaviour and properties using the principles of material science and physics. This knowledge of coating structures together with an understanding of the mechanisms of degradation processes that operate on surfaces[1-3], has allowed the development of precisely designed surfaces/coatings with enhanced degrees of corrosion resistance, wear resistance and biocompatibility.

More broadly, *Advances in Surface Engineering* provides a lens for viewing fundamental changes in the SE and corrosion and wear management professions. In this era of advanced manufacturing technologies and virtual networks, most factors of production are available globally. Capital essentially flows freely; machines can be bought or their capacity rented; technology and technological mastery are readily transferable. What increasingly sets research, and its application apart are knowledge and expertise – *intellectual*, as opposed to physical, assets.

New and more powerful ways to think about problems and actions are the prized output from the *4th International Conference on Advances in Surface Engineering*. New theories enable engineers to conceptualize their activities in novel ways and initiate more effective programmes of action. New theoretical inspiration – ranging across a broad spectrum – can also help academic researchers redefine their efforts. Particularly prized is a new methodology or way of thinking that totally transforms the shape of a field and the leveraged efforts of hundreds of researchers.

Advances in Surface Engineering is structured in 3 volumes. **Volume 1** concerns fundamental aspects of corrosion and wear. **Volume 2** gives an appreciation of SE technologies. **Volume 3** deals with applications of surface engineering to selected industrial sectors – areas central to surface engineering and holding particular promise for improvements in existing and emerging surface engineering techniques.

The *Introduction* to *Advances in Surface Engineering* aims to provoke new debate and comprehension to devise a coherent and integrated framework for tackling the enduring engineering problem of understanding and controlling corrosion and wear exploiting SE as a sustained business asset. Embedded in the *Introduction* are attempts to identify:

1. What were the important issues in SE in 1992 (the time of the *3rd International Conference on Advances in Surface Engineering*)?
2. What has happened in SE since 1992?
3. What are the important SE issues in 1997?
4. What are likely to be the important issues – particularly relating to management – in SE in 2010?

Finally, we take the opportunity to present a synopsis of the activities of the University of Northumbria's (UNN) *Surface Engineering Research Group (SERG)*.

2 SURFACE ENGINEERING

2.1 Background

The growing importance of surface engineering is due to the realization that modified, treated and coated surfaces can prevent degradation processes more effectively, particularly those which originate at surfaces[4]. This applies to a wide range of engineering applications, as exemplified by the wear coatings listed in Table 1.

It is now widely recognized – see for instance the applications cited in Table 1 – that the successful exploitation of these processes and coatings may enable the use of simpler, cheaper and more easily available substrate materials, with substantial reduction in costs, minimization of demands for strategic materials and improvement in fabricability and performance. In demanding situations where the technology becomes constrained by surface-related requirements, the use of specially developed coating systems may represent the only real possibility for exploitation[4,6–9].

Table 1 *Industries and components using thermally applied wear coatings[5]*

Aero gas turbines	Land-based turbines	Others
* Turbine and compressor blades, vanes	• Turbine and compressor buckets, vanes, nozzles	◆ Feed rolls
* Gas path seals	• Piston rings (IC engine)	◆ Pump sleeves
* Mid-span stiffeners	• Hydroelectric valves	◆ Shaft sleeves
* Z-notch tip shroud	• Boiler tubes	◆ Gate valves, seats
* Combustor and nozzle assemblies	• Wear rings	◆ Rolling element bearings
* Blade dovetails	• Gas path seals	◆ Dies and moulds
* Flap and slat tracks	• Impeller shafts	◆ Diesel engine cylinder
* Compressor stators	• Impeller pump housings	◆ Hip joint prostheses
		◆ Hydraulic press sleeves
		◆ Grinding hammers
		◆ Agricultural knives

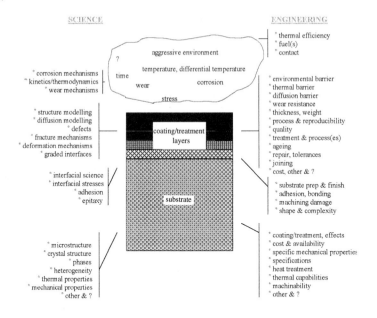

Figure 1 *Aspects of surface engineering research[1]*

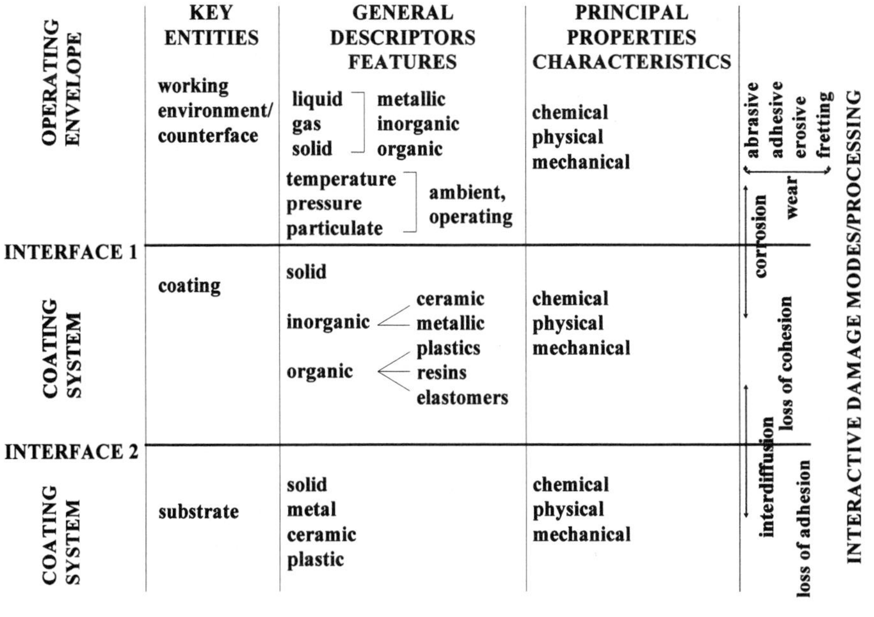

	KEY ENTITIES	GENERAL DESCRIPTORS FEATURES	PRINCIPAL PROPERTIES CHARACTERISTICS	
OPERATING ENVELOPE	working environment/ counterface	liquid ⎤ metallic gas ⎬ inorganic solid ⎦ organic temperature ⎤ ambient, pressure ⎬ operating particulate ⎦	chemical physical mechanical	abrasive / adhesive / erosive / fretting — wear — corrosion
INTERFACE 1				
COATING SYSTEM	coating	solid inorganic ⟨ ceramic / metallic organic ⟨ plastics / resins / elastomers	chemical physical mechanical	loss of cohesion
INTERFACE 2				
COATING SYSTEM	substrate	solid metal ceramic plastic	chemical physical mechanical	interdiffusion — loss of adhesion

INTERACTIVE DAMAGE MODES/PROCESSING

Figure 2 *Generalised features of a working coating system[6]*

Surface engineering produces surfaces with a unique combination of bulk and surface properties resulting in the creation of a high performance composite material. However, the biggest benefit that flows from the use of surface engineering lies in the ability to create new surfaces with highly non-equilibrium structures.

2.2 Corrosion- and Wear-Related Failures

The basic features of a simple wear-or corrosion-resistant coating system are shown in Figures 1 and 2. Refering to Figure 2, selective interaction is required at *Interface 1* to provide a wear- or corrosion-resistant surface. Selective interaction of a different kind is required at *Interface 2* to obtain adequate adhesion. Such interaction(s) must not lead to the removal of coating constituents and/or their dilution by interdiffusion across *Interface 2*. The requirement for prolonged sustainability of the corrosion- and wear-resistance of the coatings imposes additional constraints on the design of the coating system. The coatings must contain a reservoir of elements to sustain the required selective interactions at the surface (*Interface 1*). Other constraints on the coating design may flow from the necessity of a surface to resist a number of degradation processes occurring at the same time. To satisfy these requirements at both surfaces or to prevent conjoint actions of different modes of degradation, multicomponent/ multiphase and multilayered coatings are required. Even so, adequate coating systems can now be designed and produced using intelligent combinations of various processes[10,11]

2.3 Surface Engineering Technologies

Surface engineering techniques generally consist of surface treatments where the compositions/structures or the mechanical properties of the existing surface are modified, or a different material is deposited to create a new surface. Surface engineering is essential in the application and exploitation of high performance engineering components. This is especially true in relation to both the rising costs of advanced performance structural materials and the increasingly high life-cycle costs associated with high performance systems. Table 2 illustrates the market share for various types of surface finish[12].

Deposition procedures, include traditional electrodeposition and chemical conversion coating, together with thermal spraying – where a plasma or electric arc melts a powder or wire source, and droplets of molten material are sprayed on to the surface to produce a coating; PVD, in which a vapour flux is generated by evaporation, sputtering or laser ablation; and CVD, where reaction of the vapour phase species with the substrate surface produces a coating.

Surface treatments include:
* mechanical processes that work-harden the surface – e.g. shot-peening;
* thermal treatments which harden the surface by quenching constituents in solid solution – e.g. laser or electron beam heating;
* diffusion treatments which modify the surface composition – e.g. carburizing and nitriding;
* chemical treatments that remove material or change the composition by chemical reactions – e.g. etching and oxidation; and
* ion implantation – see Table 3 for applications – where the surface composition is modified by accelerating ions to high energies and implanting them in the near-surface[4].

Table 4 lists characteristics of a coating which are important in relation to quality assurance.

Table 2 *Surface finishing – value by industry sectors[12]*

Coating	Size (£M)	Share (%)
Organic	1,450	43
Plating	705	21
Galvanizing	355	10
Surface heat treatment	325	10
Hard facing	100	3
Anodizing	65	2
Tin plating	40	1
Vitreous enamelling	40	1
PVD & CVD	25	1
Others	250	8
Total	3,355	100

Comparisons of certain of the above techniques are contained within Table 5 and examples of deposition and treatment technologies used in the aerospace industry are given in Table 6.

2.4 State of the Art and Future Developments

Since the early 1980s there has been a continuing and rapid development of advanced surface engineering practices for the optimization of corrosion and wear resistance. It is now possible to produce coatings of novel composition and microstructure in multilayer/multicomponent format as appropriate to the design specification, by a variety of sophisticated physical and chemical processes, including hybrid technologies. The paradox concerning compatibility between the environment, coating and substrate is no longer a problem. At the level of research scientists and engineers, efforts must be made to enhance and systematize understanding of the various process technologies. For instance in PVD, one such issue – which offers a distinct competitive advantage – concerns plasma densities and their importance

Table 3 *Industrial exploitation of ion implantation[13]*

Material	Application (specific examples)	Typical results
Cemented WC	Drilling (printed circuit board, dental burrs etc)	Four times normal life, less frequent breakage and better end product
Ti–6Al–4V	Orthopaedic implants (artificial hip and knee joints)	Significant (400 times) lifetime increase in laboratory tests
M50, 52100 steel	Bearings (precision bearings for aircraft)	Improved protection against corrosion, sliding wear and rolling contact fatigue
Various alloys	Extrusion (spinnerets, nozzles and dies)	Four to six times normal performance
D2 steel	Punching and stamping (pellet punches for nuclear fuel, scoring dies for cans)	Three to five times normal life

Table 4 *Characteristic properties of a coating*

Structural	Mechanical	Physical	Chemical
Composition	Adhesion	Specific heat	Chemical stability
Density	Cohesion	Thermal expansion	Environmental compatibility
Porosity	Hardness		Corrosion resistance
Phase contents	Modulus		Biocompatibility
Crystallinity	Ductility		
Grain size	Strength		
Amorphosity	Fracture toughness		
Defect structures	Internal stress		
Dislocations	Wear resistance		
Vacancies	Friction coefficient		
	Deformation mode		

in facilitating the repeatable manufacture of advanced surface engineered artefacts with outstanding properties and performances[10,11]. However, a further critical problem with regard to quality assessment is the precise significance of measured properties and characteristics – e.g. hardness – in relation to actual coating performance. Here insight is needed into the consequence of particular hardness levels in relation to wear performance, so that a coating engineered to a given hardness could be expected to offer a predetermined design wear life.

At the fundamental level there is a need to understand structure/property relationships in, for example coatings, so that surface engineered systems can be designed from conception to develop desired properties. This also requires a better understanding of the degradation processes which need to be prevented/minimized by the designed surface. In this regard there is considerable scope for the creation of tailored coatings of chosen composition, structure and properties – including multilayer/multicomponent format – by highly adaptable PVD and CVD technologies (Table 7).

Table 5 *Comparison between five surface engineering processes[14]*

Process	Resistance to wear	Risk of distortion	Resistance to impact	Convenience	Range of materials
Plasma spraying – atmospheric	High	Low	Low	Very good, gun is offered to the work	Extensive
Plating	High	Low	Medium	Low, work is processed in a bath	Low
Welding	Medium	High	Good	Good	Medium
CVD/PVD ion deposition	High	Low	Good	Low vacuum chamber required	Good
Cladding	Low	Low	Good	Good	Low

Table 6 *Surface engineering technologies used in the aerospace industry[15]*

Technique	Material	Requirement	Application
Mechanical treatments, eg peening	Steels, titanium-based and nickel-based alloys	Improved mechanical and wear properties	Compressor blade roots
Paints	Phenolic and epoxy polyurethanes	Cosmetic, corrosion and wear, earthing, emissivity and infrared	Shafts, discs, blading
Polishing	Steels, titanium-based and nickel-based alloys	Cosmetic, salvage and repair efficiency	Aerofoil surfaces on vanes and blades
Electrochemical	Tribomet, chromium	Corrosion and wear, salvage and repair	Bearing chambers, stator vanes
Thermal spraying (D-gun, flame spraying, plasma spraying)	Al/Si polyester, WC/Co, CuNiIn	Corrosion and wear, salvage and repair, seals, net-shapes	Snubbers, gas-path seals, combustor cans
Thermochemical	Nitrogen and carbon into steels	Improved mechanical properties	Shafts and gears
Pack aluminizing	Nickel-based alloys	Corrosion/oxidation	Aerofoil surfaces on vanes and blades

2.5 Quality Assurance

Quality assurance of surface engineered coatings and surfaces is a major issue particularly for the coating users and producers[10].

In the absence of a definitive knowledge of the structure/property relationships in coatings deposited on a surface, only an empirical approach can be adopted[10]. A coating can be described in terms of its characteristic properties (Table 4).

One approach which is being increasingly adopted is to define the functionality of the coating for a particular application in terms of a sub-set of the properties listed above. For example, to achieve quality assurance of a load-bearing prosthesis, consideration can be given, in the first instance, to the sub-set of properties consisting of adhesion, strength, wear resistance, friction and biocompatibility. Similarly corrosion resistant coatings can be quality assured by addressing parameters such as adhesion, residual stress, ductility, K_{Ic}, fatigue/crack growth resistance and chemical stability.

Quality assurance of deposition technology[17], as well as manufactured and surface engineered artefacts to reliably impart specified measurable performances, is central to surface engineering. Effort needs to be made to develop expert systems to design and select coating/treatment systems and hence define the appropriate process technology.

These skills and competencies must be applied within the construct of business realities – achieving sustained world-class competitive advantage.

Table 7 *Future surface engineering activities*[4,16]

---

1. Surface engineering of non-ferrous metals
2. Surface engineering of polymers and composites
3. Surface engineering of ceramics
4. Mathematical modelling of surface engineered components
5. Surface engineering in material manufacture
6. Statistical process control in surface engineering
7. Non-destructive evaluation of surface engineered components
8. Duplex or hybrid surface engineering technologies and design, eg:
   * laser treatment of thermal and plasma spray coatings
   * ion beam mixing and ion-assisted coatings
   * hot isostatic pressing of overlay coatings
   * thermochemical treatment of pre-carburized steels
   * thermochemical treatment of pre-laser hardened steels
   * CVD treatment of pre-carburized steels
   * PVD treatment of pre-nitrided steels
   * ion implantation of pre-nitrided steels

---

3 MANAGEMENT ISSUES

Hard and anecdotal findings from research into the strategic management of technology and the exploitation of technological innovation suggest that, for leading firms in a wide variety of industries, developing advanced technologies *per se* is rarely the constraining challenge in technological innovation. Rather, the challenge is innovation in the market. In this regard quality management, management of change, strategic management, innovation and knowledge management are as important as technological issues.

3.1 Quality

Bench-marking is a means of measuring and comparing performance and may be defined as, "the continuous process of measuring products, services and practices against the toughest competitors or those companies recognized as industry leaders". The emphasis should be on understanding how surface technologists carry out their activities; learning how other groups excel in surface engineering; and then adapting and reinterpreting what has been learnt in a way that makes for competitive advantage. Instead of aiming to improve only against previous performance, technologists should use bench-marking to inject an element of imagination into the quest for progress, while simultaneously and objectively scrutinizing established processes. Inevitably, the perspective is international[18,19].

As part of the normal research work an external audit could be carried out, this may comprise a survey of the controls surrounding researchers' daily records – before and after implementation of quality improvements. Assessments of the following areas could be among the audit's key functions:
* assurance of daily task completion;
* periodic review by management;
* verifiability of results claimed against physical evidence;
* evidence of consistency/quality of results claimed; and
* indication of appropriate coordination with clients.

Items for further investigation might also include:
- How is control exercised by management – formally or informally?
- Is control proportional to risk, exposure and objectives?
- Are guidelines for the delegation of authority suitable for the organization?

A standard part of the audit would be an analysis of the effectiveness of researchers' time management. The *day in the life of* technique consists of unobtrusively observing the activities of selected researchers over a period of several days. The allotment of their – *value added* or *non-value added* – time should be reviewed with both the researcher and manager. For *non-value added* activities the root causes of the operation would be investigated and minimized[20].

3.2 Change

Management of strategic and organizational change must address three key questions:
1. How is change initiated and implemented in relatively successful organizations?
2. What is the rôle of the management at the service provider and end-user companies in initiating change?
3. What is the contribution of management development in the implementation of organizational and strategic change?

Change management must at the same time stimulate innovation and provide mechanisms for dealing with uncertainty in knowledge-intensive SE business environments. This might be accomplished by systematically analyzing the organization's values, needs, interests and relationships, and productively applying the insights gained. An important aim of this process is to determine – using multiple levels of comparison – specific factors that might lead to competitive advantage.

Individual organizations and consortia need to assess complementary plant networks – here the objective should be to provide qualitative insight and analytical tools to facilitate the development and adaptation of plant networks in response to diverse and changing markets, manufacturing costs and technological capabilities across countries[21].

3.3 Strategy

During times of tumult strategic alliances become increasingly attractive. Research has indicated that to make strategic alliances work, success does not necessarily come from the structural or systems aspects of the alliance, but rather from the quality of relationships, the degree of trust, mutual commitment and the flexibility of attitude brought to the relationship. General assessments of the management challenges associated with creating and managing strategic alliances must consider the dynamics of cross-functional partnerships and a specific evaluation of the rôle of international alliances in high-tech SE industries. Alliances need to be analyzed from three perspectives:
a. direct economic costs and benefits;
b. historical evolution; and
c. external networks.

Growing reliance on strategic alliances has prompted investigations into how alliance partners utilize complementary technological capabilities to create products that neither could develop individually and what drives companies to choose strategic alliances and joint ventures over internal R&D and licensing as approaches to augmenting technological capabilities. Also being explored are how technology-based companies develop and exploit strategically valuable knowledge assets and the managerial and technical processes by which novel technologies are applied to the development of complex new products. Other studies are suggesting ways in which senior executives can nurture superior product development performance and advancing the notion that technological innovation is often less a constraint than the need to innovate in the marketplace. Likewise, research suggests the principal challenge in entrepreneurship is building skills such as selling ideas and products and applying theories of competitive advantage, not communicating new knowledge or developing new theories that better explain entrepreneurship. Of particular interest is how technological knowledge is transferred across company boundaries and the rôle collaborative development might play in countering a company's core rigidities, i.e. deeply in-grained, but out-dated, technological capabilities.

Strategic decision-making may be considered as essentially a technique for making judgements when the outcome depends in part on the actions of others, it involves systematic analysis consistent with the commonplace expedient of putting oneself in another's shoes, and interactive decision and value analysis. The process may involve clinical, statistical and theoretical research focused on major commitments, notably investments and disinvestments.

The impact of technology on industry structure – such as investments in SE technologies – need to be evaluated as part of coherent business strategies and be viewed as strategic necessities rather than attempts to gain sustainable advantage. In considering technological change and competitive strategy it is necessary to explore the dynamic links between and strategic consequences of technological change and shifts in organizational structure and competitive advantage. Research attempts to answer questions such as:

* How does a firm identify opportunities to create value from technological change and substitution?
* How can firms create sustainable profits in the face of SE technological change and resulting adjustments in competitive dynamics?
* What methods are available to managers for evaluating highly uncertain projects and the value of developing specific competencies and capabilities?
* How might competitors' capabilities be assessed?
* How does technological change affect competitive dynamics and redefine industry structures?
* Under what conditions should a firm invest in a new technology?
* How do competitive dynamics affect the evolution of technology and attendant quality standards?

The merging of computers, telecommunications and SE – and the blurring of functional and technological boundaries within the SE industry – forces managers to pursue a wide range of topics, including the competitive dynamics within and between quality standards, the rôle of alliances in SE technology, and the relationship between technology choices, scope of a firm (i.e. degree of horizontal and vertical integration) and financial decisions.

Research might profitably be conducted into the determinants of superior SE process development performance and process development strategies within, for instance, the aerospace and automotive industries. Detailed qualitative and quantitative data could be collected on the histories, strategies and performance of surface engineered artefacts. Statistical analysis could then be used to identify how such factors as organizational structure, project strategy, organizational capabilities and technological environment influence process development lead times, productivity and costs. Such a study would be expected to shed light on the special challenges that attend the management of R&D projects and building of development capabilities in SE-based industries; in addition this would yield significant insights into the potential strategic rôle of SE technologies in these important industries over the coming decade[21].

3.4 Innovation

Innovation – the successful exploitation of new ideas – is the key to sustained competitiveness. British university research groups have responded to the Government's "Science, Engineering and Technology" and "Competitiveness" White Papers, and especially that aspect which is concerned with the application of new techniques and ways of working that improve the effectiveness of individuals and organizations. The scope of several research programmes covers the rôle of innovative management in the achievement of sustained improvement in the bottom-line performance of commercial and industrial businesses. In particular emphasis is placed on the human and organizational processes and conditions that contribute to this. Such "research on business in business" is designed to increase knowledge and understanding of these crucial elements and to encourage their exploitation by industry[22].

Organizations frequently manage innovation and development activities not only as single projects but also as a cohesive set of related projects. Research is currently underway in a number of organizations to compare the different methods used for managing project sets. The aim is to identify "best practice" methods and to examine the link between the strategic aims of the organization and the formation of the goals and objectives of the projects[23].

3.5 Knowledge

In new lean, business processes where non-value added tasks have been eliminated, IT (information technology) can facilitate manufacturing philosophies, e.g. JIT (just in time). However, complex problem areas still exist in re-engineering companies. These problems will occur with large and small companies, and basic IT cannot address them. They include the handling of incomplete, conflicting and vague data, the discovery of knowledge in massive data sets, the interpretation of legislation and inter-organizational contracts, the management of change, and the re-application of an expert's accrued experience and expertise. *Knowledge-based systems* (KBSs) provide a series of techniques that can help to assess, manage and ameliorate these problems. There have been a number of reported successes where KBSs have added value to business processes, and in fact made business re-engineering possible[24].

A *knowledge-based system* refers to any assemblage which incorporates a level of expertise and experience, which can be used to address new situations in an intelligent way. The knowledge may derive from human *experts*, research papers and reports, or computer systems. KBS technology – e.g. expert systems, neural networks, data mining and artificial intelligence – applied to SE can offer industry significant benefits:
- intelligent decision-making support, making safer, faster and more effective decisions;
- consistency of approach and assurance of quality standards;
- better service delivery, increased productivity and improved cost control;
- dissemination of scarce expertise across the organization;
- a valuable training tool for new engineers and managers; and
- developing a way of making complex situations more transparent for the decision-maker and linking shop-floor activity to commercial transactions.

The technical focus might be concentrated in the fields of knowledge-based SE and corrosion and wear management, and it is in these areas that developed solutions to complex industrial problems should be sought[25].

4 THE *SURFACE ENGINEERING RESEARCH GROUP*

4.1 Background

The *Surface Engineering Research Group* (*SERG*) comprises a Director (Prof. P. K. Datta), an Assistant Director (Dr J. S. Burnell-Gray), 5 academic consultants, 2 Research Fellows, 2 Research Associates and 4 Research Students. It has five core functions: *Research, Education, Technology Transfer, Consultancy* and *Training*.

SERG firmly believes that surface engineering is one of the several keys to UK industry gaining a world-class competitive advantage. The *Group's* objectives are to identify outstanding pivotal research issues, promote the take-up of SE research results in industry and influence UK and European Union research policy to help create the next generation of manufacturing systems.

Our technical focus is concentrated in the fields of surface engineering, corrosion and wear, and it is in these areas that we seek to develop commercial applications as solutions to complex industrial problems. The research focus of the laboratory is addressed along with partners in for instance Rolls-Royce, Multi-Arc (UK), Chromalloy UK and Johnson Matthey. *SERG's* research portfolio also supports cross-fertilization of the work of other groups.

The development of successful applications in corrosion and wear management is not a straightforward task as it requires the synergistic combination of expertise on coatings deposition, surface engineering, corrosion and wear engineering, as well as corrosion/wear theory. However, the rewards of SE are across-the-board improvements in value, quality, customer support and productivity.

4.2 Research Portfolio

It is within this demanding and continually changing framework – also see Figure 3 – that the *Group's* research activities are based. Added-value to the component in relation to enhanced mechanical, thermal, chemical, electrical or optoelectrical attributes, as well as fitness-for-purpose and the minimization of life cycle costs, are the driving forces for the justification and adoption of surface engineering practices. In this regard the *Group's* research has not only contributed to the characterization and understanding of the functional behaviour of coatings, but also at a more fundamental level has involved the systematic application and testing of scientific principles with a view to creating entirely new types of surfaces with novel properties. More pragmatically this research effort has in addition contributed to the achievement of more reliable coatings with reproducible properties. This reproducibility aids prediction of the performance, e.g. mechanisms and time-dependent behaviour, of surface-modified artefacts using surface analytical techniques, system modelling, interfacial simulation and NDE – all of which are essential if engineers are to fully exploit the potentials of the discipline.

The scope of research programmes – from basic, through strategic/pre-competitive to applied – is outlined in Table 8. Other examples of applied surface engineering research lie not only in the field of aero engine turbine blades, but also in natural gas production and combustion, and orthopaedics.

The *Surface Engineering Research Group* advocates the principles of surface engineering and provides local and national industry with a world-class facility for the study of corrosion and wear control using surface engineering technologies. Since the early 1980s *SERG* has gained an international reputation in the area of surface engineering and also acts as a regional teaching facility in corrosion and wear prevention using the latest surface engineering technologies involving surface analysis, surface modification and coatings deposition. *SERG* offers a well-founded mechanical engineering workshop and corrosion laboratory comprising computer-aided machining, non-traditional machining, and high temperature gaseous and molten salt, as well as aqueous corrosion facilities.

Table 8 *SERG's research portfolio*

Current and Recent Programmes

- *Design and Optimization of High Temperature (HT) Protective Coatings* concerns the design and development of HT degradation resistant MCrAlYX-type coating systems capable of withstanding corrosion in coal gasifier atmospheres typically containing significant Cl_2 or S_2 potentials. The project formed part of the EPSRC Rolling Programme in Surface Engineering jointly pursued by ourselves, Hull University and Sheffield Hallam University.

- *Interfacial Modification of MMCs* aiming to improve HT mechanical properties and corrosion resistance of Ti-Ti alloy matrix/SiC fibres. Interfacial modelling is used to quantitatively and qualitatively describe diffusion and corrosion mechanisms. HT studies are performed in atmospheres designed to simulate aero-engine compressors. Part of the EPSRC Rolling Programme with UNN support.

- *Optimization of Electroless Deposition Processes for Protective Coatings* studying and exploiting selected parameters (bath chemistry, and coating composition and morphology) of electroless Ni-B and Ni-P coatings with varying concentrations of B and P, as a function of coating condition and determining the resultant effect on corrosion, wear and fatigue behaviour. Part of the EPSRC Rolling Programme.

- *Design and Development of Pt-Aluminide Coatings for Improved Corrosion Resistance* determining a reference database and enhancing Pt-modified aluminide coatings, deposited with or without Ta, on superalloy substrates designed for use as gas turbine blades. Hot corrosion is monitored in simulated gas turbine environments. Sponsored by EPSRC, Rolls-Royce plc, C-UK Ltd. and Johnson Matthey.

- *Improved High Temperature Resistant Silicon Nitride-Silicon Carbide Composites* concerning the characterization and optimization of Si_3N_4-SiC composites during exposure to replicated combustion environments. The EU provided funds for this research programme, jointly pursued by ourselves, Limerick University, T&N Technology and British Gas.

- *High Temperature Corrosion of Car Engine Valve and Valve Seat Materials* involving the study and selection of appropriate Cr_2O_3- and Al_2O_3-forming alloys, oxide dispersion-containing and other mechanically alloyed materials, to optimize durability in simulated car engine atmospheres. Funding by EPSRC and British Gas.

- *Surface Engineered (Diamond-Like Carbon) Prostheses* embraces a wide range of activities, viz: depositing and characterizing DLC coatings, biocompatibility tests, and fatigue and wear studies of coated prostheses. Funding is from DTI Link. Partners include 3M, Teer Coatings and the Royal Victoria Infirmary (Newcastle).

- *Studies of Electroless Coatings* concerning the deposition and characterization of electroless coatings on a number of substrates. The aim was to produce coatings with a set of specific properties such as high corrosion resistance, texture, appearance and good adhesion. The project was sponsored by NPL.

Completed Programmes

* *Coatings Technology*
  - Development of Ta and Nb ion-plated coatings as novel load-bearing (human) implant materials

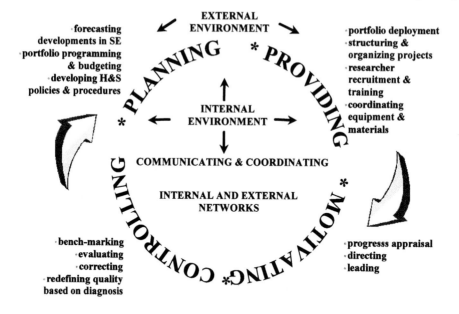

Figure 3 *SERG's research management functions*

- Electrolytic deposition of Co-Sn alloys for the electrical industry
* *Aqueous Corrosion*
  - Corrosion of implant systems in bodily fluids
  - Corrosion of electroless coating systems
  - Passivation in electrolytically deposited alloys
* *Environmental Cracking*
  - Stress corrosion cracking (SCC) in titanium alloys used in submarine hulls
  - Development of software and hardware to study and analyse crack propagation
  - SCC of nodular cast irons in heavy goods vehicle (HGV) suspensions
* *High Temperature Degradation*
  - Chloridation of binary alloys, MCrAlX-type and MCrAlY-type coatings alloys
  - Sulphidation/oxidation of advanced engineering ceramics, Si_3N_4/SiC composites, CoCrAlYX-type coatings alloys, refractory metals, HfN and Nb_2N
  - Oxidation of titanium alloys and MMCs for aerospace applications

---

4.3 Future Challenges, Development and Plans

Plans to consolidate and expand on the success enjoyed by *SERG* centre on:

Research   ◆ Expanding into interfacial modelling and interfacial engineering;
                    ◆ moving into the area of pack cementation;
                    ◆ diversifying into deposition (PVD) control and optimization;
                    ◆ assessing the viability of PVD catalyse;

> ◆ moving into new pre-treatments, e.g. pre-nitridation and pre-carburization;
> ◆ extending electroless deposition into cobalt alloys;
> ◆ moving away from sulphidation towards chloridation and erosion/corrosion;
> ◆ introducing the concept of knowledge-based systems;
> ◆ contingency funding to refurbish/maintain existing equipment;
> ◆ buying a mass spectrometer;
> ◆ constructing a burner rig; and
> ◆ further exploiting IPR.

Education
> ◆ Offering the facility of split PhDs, interchange of research fellows and travelling professors;
> ◆ providing a surface engineering module to existing postgraduate and EPSRC IGDS courses; and
> ◆ developing an IGDS in Research Management.

Technology Transfer
> ◆ Presenting the 5th International Conference on Advances in Surface Engineering in 2000;
> ◆ hosting the 1998 Institute of Corrosion's Corrosion Science Symposium;
> ◆ hosting the 2nd European Workshop in Surface Engineering Technologies for SMEs in 1998;
> ◆ promoting in-house courses in corrosion, wear and surface engineering based on EPSRC's Rolling Programme in Surface Engineering commencing 1998; and
> ◆ further enhancing consultancy and contractual services to industry.

References

1. J. S. Burnell-Gray and P. K. Datta (editors), 'Surface Engineering Casebook' Woodhead Publishing, November 1996.
2. K. N. Strafford, P. K. Datta and J. S. Gray (editors), 'Surface Engineering Practice: Processes, Fundamentals and Applications in Corrosion and Wear', Published by Ellis Horwood, 1990.
3. P. K. Datta and J. S. Gray (editors), Conf Proc 3rd Int Conf Advances in Surface Engineering, Newcastle upon Tyne, May 1992. Published by The Royal Society of Chemistry, 1993, 'Surface Engineering Vol I: Fundamentals of Coatings', 'Surface Engineering Vol II: Engineering Applications', 'Surface Engineering Vol III: Process Technology and Surface Analysis'.
4. V. Sankaran, 'Surface Engineering – A Consultancy Report', Advances in Materials Technology: Monitor, Issue 24/25, February 1992.
5. P. Sahoo, *Powder Metallurgy International*, 1993, **25**, 73.
6. K. N. Strafford and S. Subramanian, *J. Materials Processing Technology*, 1995, **53**, 393.
7. Reference 2, Keynote paper, Chapter 3.1.1, p. 397.
8. J. S. Burnell-Gray and P. K. Datta (editors), 'Quality Issues in Surface Engineering', To be published by Woodhead Publishing 1997.
9. J. S. Burnell-Gray, Internal reports, MBA course, University of Sunderland , 1995-1997.

10. Correspondence with K. N. Strafford, University of South Australia.
11. Correspondence with K. N. Strafford, University of South Australia.
12. D. Hemsley, *Engineering*, 1994, **235**, 25.
13. P. Sioshansi, *Thin Solid Films*, 1984, **118**, 61.
14. I. H. Hoff, *Welding and Metal Fabrication*, 1995, **63**, 266.
15. D. S. Rickerby and A. Matthews, 'Advanced Surface Coatings: A Handbook of Surface Engineering', Published by Blackie, 1991.
16. T. Bell, *J. Phys D: Appl Phys*, 1992, **25**, A297.
17. C. Subramanian, K. N. Strafford, T. P. Wilks, L. P. Ward and W. McMillan, *Surface and Coatings Technology*, 1993, **62**, 529.
18. A. van de Vliet, *Management Today*, January 1996, 56.
19. F. C. Allan, *Special Libraries*, 1993, **84**, 123.
20. S. J. Burns, *Internal Auditor*, 1991, **48**, 56.
21. Anon, Harvard Business School, Internet home pages: www.hbs.edu/research/summaries/lec.html, accessed 21st October 1996.
22. Anon, Economic & Social Research Council, Internet home pages: www.bus.ed.ac.uk:8080/ESRC-Innovation.html, accessed 23rd October 1996.
23. Anon, Cranfield University, School of Management, Internet home pages: www.cranfield.ac.uk/som/res/default.html, accessed 18th October 1996.
24. Anon, Ulster University, Northern Ireland Knowledge Engineering Laboratory, Internet home pages: www.nikel.infj.ulst.ac.uk/nintro.htm, accessed 23rd October 1996.
25. Anon, London University, London Business School, Internet home pages: http://www.lbs.lon.ac.uk/om/research.html#tech: accessed 18th October 1996.

Section 1.1 High Temperature Corrosion

1.1.1
The Oxidation Behaviour of Lanthanum Implanted Stainless Steels

F. J. Ager[1], M. A. Respaldiza[1], A. Paúl[2], J. A. Odriozola[2], C. Luna[3], J. Botella[3], J. C. Soares[4] and M. F. da Silva[5].

[1]DEPARTAMENTO DE FÍSICA ATÓMICA, MOLECULAR Y NUCLEAR. UNIVERSIDAD DE SEVILLA. SPAIN
[2]UNIVERSIDAD DE SEVILLA-CSIC. SPAIN
[3]ACERINOX S.A., LOS BARRIOS, CÁDIZ, SPAIN
[4]CFNUL, AVDA. PROF. GAMA PINTO 2, P-1699 LISBOA CODEX, PORTUGAL
[5]INSTITUTO TECNOLOGICO E NUCLEAR, ESTRADA NACIONAL 10, P-2685 SACAVÉM, PORTUGAL

1 INTRODUCTION

Rare earth oxide deposition onto stainless steel surfaces has been attempted as a way of improving corrosion resistance at elevated temperatures[1-4]. The improvement in the corrosion behaviour has been related to the modification of the diffusion mechanisms through the chromia protective layer. In a previous work we have postulated the formation of a $LaCrO_3$ as responsible for such a behaviour[5]. Among the alternatives to deposit reactive elements, ion implantation has been chosen as a way of obtaining surface and/or subsurface alloys with the desired composition[6]. During ion implantation, a modification of the alloy structure may also occur, resulting in a way of testing the influence of the alloy structure on the oxidation behaviour.

In the present work we propose two procedures for obtaining the refractory behaviour: implantation in the bulk alloy and in controlled preoxidized layers. Ion fluency has been chosen in such a way that final rare earth element concentration falls within the limits experimentally observed as adequate using wet chemistry methods.

Excellent parabolic oxidation is observed in every case showing the efficiency of the implantation method both in the implanted bulk alloy as well as in the preoxidized specimens. The differences in the oxidation kinetics are related to the surface composition and to the structure of the implanted materials.

2 EXPERIMENTAL PROCEDURE

The oxidation behaviour of AISI-304 stainless steel (18Cr/8Ni) samples was studied after the implantation of lanthanum ions at several doses, and compared with the oxidation of the conventional unimplanted steel. The size of the samples was 10x10x2 mm, resulting in a total surface of 2.8 cm^2 (faces and edges). Two surface finishings have been tested: surface preoxidation in air at 900°C for two minutes and commercial 2B finishing.

Oxidation of prepared specimens at 900°C was performed in situ in a SETARAM TGDTA-92 thermobalance, for thermogravimetric (TG) measurements, with a sensitivity of 0.01 mg and with a temperature control of ±1°C, being the temperature increase rate of β=100 K min^{-1}. At 900°C, the AISI-304 has no longer a refractory behaviour. The oxidation process consisted of heating the samples in running synthetic air (1 bar), up to 900°C and keeping them at that temperature for 1 hour, followed by furnace free cooling to room temperature in synthetic air (1 bar).

The implantations were performed in a 200 kV Danfysik S1090 ion implanter, using 100 keV La^+ ions, doses of 0.8×10^{16}, 10^{16} and 1.2×10^{16} at/cm². The samples were implanted in one side at a time, and each implantation was controlled with an aluminium sample, since no sputtering effects are expected at these doses for aluminium. The edges or lateral sides were not implanted. Table 1 summarizes the analysed samples and their treatments.

Rutherford backscattering spectrometry (RBS) characterization of the oxide layers was done with the 3.1 MV Van de Graaff accelerator at Sacavém, using 1.6 MeV protons and alpha particles. The incident beam, normal to the sample surface, was collimated up to about 1 mm diameter. Backscattered particles were detected at 140° and 180° with silicon surface barrier detectors. The solid angles and integrated charge are known with an uncertainty of about 3%. The spectra were analysed using the RUMP computer program[7] with modified cross-sections for the non-Rutherford scattering of 1.6 MeV protons by oxygen. The implanted doses were also measured with this technique and agree, within the experimental errors, with the nominal doses shown in Table 1.

The surface of the specimens was analysed by optical microscopy and grazing incidence FT-IR.

XRD patterns were obtained from a Siemens Krystalloflex D-5000 spectrometer using Cu-K_α radiation and pyrolitic graphite as monochromator. SEM experiments were performed on a Philips XL30 (30 kV), with a silicon surface barrier detector for backscattered electrons and LiF photomultiplier for secondary electrons. For the EDX analysis, a SiLi detector with Be window and 10 mm² of active area was used.

3 RESULTS AND DISCUSSION

3.1 Modifications Induced by the Implantation

The first visible effect of the implantation upon the AISI-304 specimens is a change in the crystalline structure of the surface. SEM images (Figure 1) show a recrystallization of the surface after the implantation with lanthanum, and the appearance of a new grain structure on top of previous grains. This fact must be related to the appearance of a bcc-like phase detected by means of X-ray diffraction at an incident angle of 0.1° (grazing incidence) (Figure 2), since the only phase present before implantation was the fcc.

Table 1 *Analysed samples, thermal treatments and nominal doses*

Sample No	Preoxidation	100 keV La^+ dose (10^{16} at/cm²)
116	2 minutes at 900°C	0.8
121	None	0.8
118	2 minutes at 900°C	1.0
123	None	1.0
120	2 minutes at 900°C	1.2
125	None	1.2

Table 2 *Thermogravimetric results showing the weight increment and the weight increment per surface unit of the samples*

Sample No	ΔW (mg)	$\Delta W/S$ (mg/cm^2)
Unimplanted	0.17 ± 0.01	0.061 ± 0.004
116	0.13 ± 0.01	0.046 ± 0.004
121	0.10 ± 0.01	0.036 ± 0.004
118	0.10 ± 0.01	0.036 ± 0.004
123	0.09 ± 0.01	0.032 ± 0.004
120	0.10 ± 0.01	0.036 ± 0.004
125	0.09 ± 0.01	0.032 ± 0.004

3.2 Study of the Oxidation Behaviour

The implantation of lanthanum at these doses has also an important effect upon the oxidation kinetics of the AISI-304 after 1 hour, as can be immediately concluded from the mass gains obtained by thermogravimetric measurements (Table 2). The unimplanted sample oxidizes clearly faster than the rest. The other samples gain roughly the same mass. We must have in mind the effect of the oxidation of the unimplanted lateral sides (30% of the total surface). Therefore we cannot expect to be sensitive to slight differences in the oxidation process due to little differences in the implanted dose. For this purpose, we must directly analyse only the oxide layer of the implanted surfaces using the RBS technique.

Figure 3 shows typical RBS spectra obtained from sample 116 (preoxidized) with 1.6 MeV protons. The oxygen signal is clearly visible in both spectra on the top of the continuous RBS spectrum from the steel matrix (mostly iron). The straight integration of the oxygen

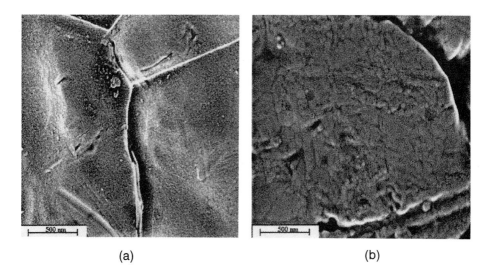

(a)   (b)

Figure 1 *SEM images of AISI-304, 2B finish, (a) before and (b) after implantation with lanthanum*

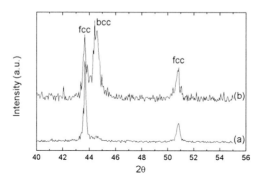

Figure 2 *XRD diffraction patterns at grazing incidence (0.1°) of AISI–304, 2B finishing, (a) before and (b) after implantation of lanthanum. We see the appearance of a bcc phase after the implantation*

peak leads to the precise amount of O diffused into the oxide layer. Besides, if we assume that most of the oxide layer is composed of Cr_2O_3, an approximation for the thickness can be computed. Table 3 summarizes these results. For each implanted dose, two samples were analysed before and after 1 hour oxidation at 900°C respectively. The analysis of the unimplanted sample is also included for comparison. We see again the difference in the oxide scale thickness grown by oxidation between the unimplanted and the implanted samples, only after 1 hour oxidation. We expect this difference to increase as we increase the oxidation time, as it was

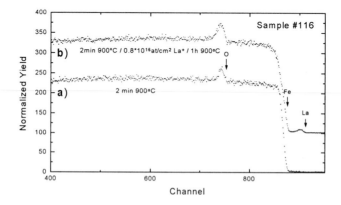

Figure 3 *RBS spectra obtained with 1.6 MeV proton scattered at 140° from sample 116 (preoxidized): (a) unimplanted; (b) after implantation and oxidation. The base line has been shifted up 100 counts/keV/μC/msr*

Table 3 *Equivalent oxide layer thickness (see text) as computed from RBS analysis with 1.6 MeV protons. Last column shows the increase in the oxide scale thickness, i.e. the increase due only to the oxidation process (1 hour at 900°C)*

Sample No	La dose (10^{16} at/cm^2)	Equivalent thickness of O (10^{16} at/cm^2)		Δ Thickness (Å of Cr$_2$O$_3$)
		Before 1h at 900°C	After 1h at 900°C	
Unimplanted	0	0	200	3200
116	0.8	77	130	850
121	0.8	0	90	1450
118	1.0	85	167	1330
123	1.0	0	110	1780
120	1.2	70	157	1410
125	1.2	0	134	2160

found for the case of coating of AISI-304 with lanthanum nitrate solutions[5] instead of implantation. Representing the data as in Figure 4, it is clear that the samples implanted with higher doses oxidize slightly faster than those implanted with lower doses. Besides, the oxide formed before implantation seems to play a protective rôle against oxidation, slowing the process, in agreement with the behaviour found when depositing lanthanum[8].

Further RBS analyses with 1.6 MeV ^{4}He$^+$ were carried out for a more detailed depth profiling of the elements involved in the oxidation and diffusion processes. Figure 5 depicts the spectra of sample 116. In this case, we can perfectly profile lanthanum since its signal is quite separated from that of iron, chromium or manganese. However, we cannot profile the latter three because their signals are too nearby to each other to be able to single them out. Only on surface we are able to distinguish Cr from Mn+Fe, and then we can assume a slow

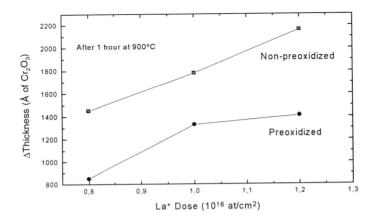

Figure 4 *Evolution of the oxide thickness with the implanted dose and the preoxidation treatment*

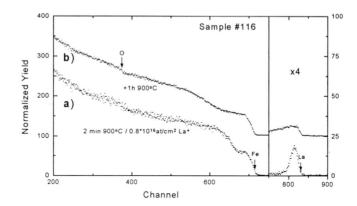

Figure 5 *1.6 MeV ⁴He⁺ spectra of sample 116 (preoxidized): (a) as implanted; (b) implantation followed by oxidation (1 hour at 900°C)*

variation in the composition until reaching that of the bulk. This is the way the RUMP computer simulations have been performed. Those species can be profiled by Nuclear Reaction Analysis (NRA), using the appropriate reactions, such as $^{54}Cr(p,\gamma)^{55}Mn$ for chromium profiling, but this study is still under progress.

Figures 6 through 9 plot the depth profiles of lanthanum and oxygen within the oxide layer for all the implanted samples. If lanthanum was implanted in the first 80 Å (non-preoxidized samples) or 200 Å (preoxidized samples) after 1 hour of oxidation it distributes itself along

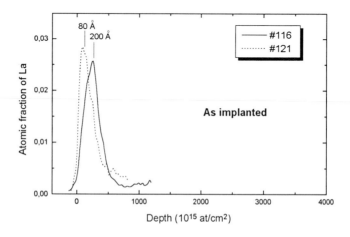

Figure 6 *Depth profiles of lanthanum "as implanted" for samples 116 (2 minutes 900°C+0.8×10¹⁶ at/cm² La⁺ (100 keV) and 121 (0.8×10¹⁶ at/cm² La⁺ (100 keV))*

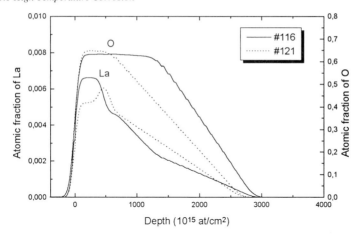

Figure 7 *Depth profiles of lanthanum and oxygen after the whole thermal treatment for samples 116 (2 minutes 900ºC+0.8×10¹⁶ at/cm² La⁺ (100 keV) +1 hour 900ºC) and 121 (0.8×10¹⁶ at/cm² La⁺ (100 keV) +1 hour 900ºC)*

the oxide layer, presenting a surface composition of about 0.6 atomic%, and then slowly decreasing towards the inner part of the scale. There are no signs of lanthanum entering the matrix beyond the oxide. Beside this, the profiles are flatter on the surface for preoxidized samples, presenting a peak below it the non-preoxidized ones. The protective effect of lanthanum can be related to its oxophilicity, which tends to form a mixed oxide on the steel surface.

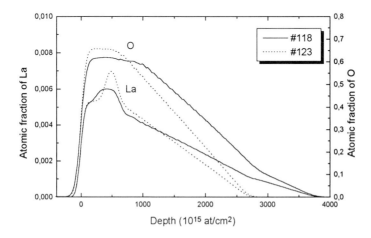

Figure 8 *Depth profiles of lanthanum and oxygen for samples 118 (2 minutes 900ºC+10¹⁶ at/cm² La⁺ (100 keV)+ 1 hour 900ºC) and 123 (10¹⁶ at/cm² La⁺ (100 keV)+1 hour 900ºC)*

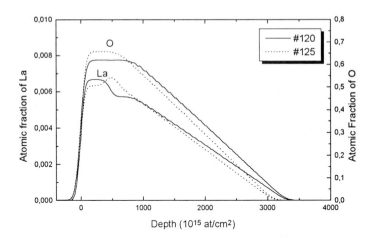

Figure 9 *Depth profiles of lanthanum and oxygen for samples 120 (2 minutes 900°C+1.2×10^16 at/cm² La+ (100 keV)+ 1 hour 900°C) and 125 (1.2×10^16 at/cm² La+ (100 keV) +1 hour 900°C)*

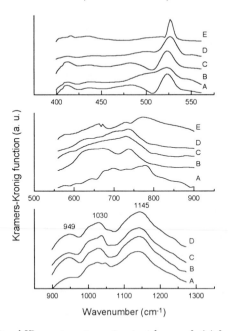

Figure 10 *Normalized IR spectra at grazing incidence of: (a) La-implanted AISI 304 sample, (b) La-implanted AISI 304 sample after 1 hour oxidation in air at 900°C, (c) La-implanted on preoxidized 2 minutes at 900°C in air AISI 304 sample, (d) La-implanted on preoxidized 2 minutes at 900°C in air AISI 304 sample and further oxidized for 1 hour at 900°C, and (e) AISI 304 sample oxidized 2 minutes in air at 900°C*

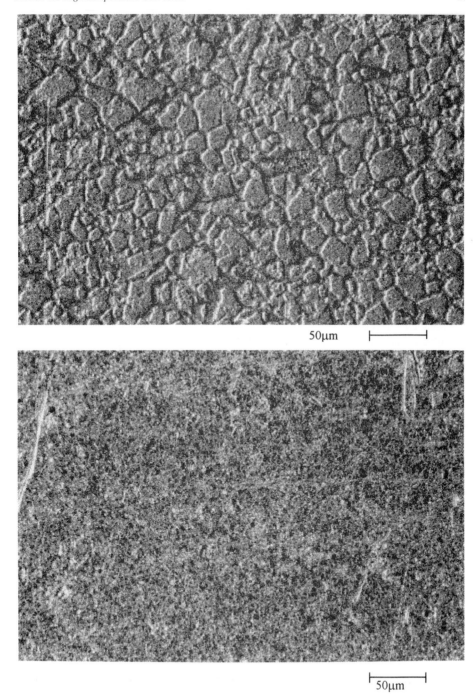

50μm

50μm

Figure 11 *Optical microscopy pictures of an unimplanted sample after 2 minutes at 900°C and before implantation (top) and after the implantation of 10^{16} at/cm^2 of 100 keV La$^+$ (bottom)*

Figure 10 shows the IR spectra taken at an angle of 87° off normal. The spectra are shown after transforming the original data using the Kramers–Kronig algorithm built on the Nicolet 510P software. Three main regions are depicted due to the presence of first derivative-like peaks resulting from specular reflections. Previous studies have characterized the initial stages of the oxidation of AISI-304 stainless steels showing that this technique is able to identify several phases formed on the surface of these materials.

Several studies have been reported on the IR reflectivity of oxide layers. Ottesen[9] studied the IR spectra of the oxide scale formed on the surface of binary Fe-Cr alloys finding bands at 400, 600, 735 and 530 cm^{-1} ascribed to Cr_2O_3. Oxide scale analysis formed on stainless steels (AISI–304) was performed by Lenglet et al[10]. They found the presence of Cr_2O_3, $MnCr_2O_4$, Fe_2O_3 and Fe_3O_4 in the initial stages of the oxidation assigning peaks at 745–730 cm^{-1} to Cr_2O_3, 690, 590 and 470–460 cm^{-1} to chromites, and 510 and 430–420 for α-Fe_2O_3. More recently, Music et al.[11] studied the structural and chemical properties of the mixed metal oxides $(Fe_{1-x}Cr_x)_2O_3$ using among others IR spectroscopy. They found that the relative ratio Fe/Cr causes a shift of the infrared bands to higher wavenumbers on increasing the chromium content. Similarly they found that sharp and intense bands at 445 and 414 cm^{-1} have to be ascribed to α-Cr_2O_3.

Two minutes preoxidation of the stainless steel results in peaks at 416, 435, 527, 586, 667, 731 and 774 cm^{-1} (Figure 10, e). According to Le Calvar and Lenglet[12] thin films (10 nm) of Cr_2O_3 on chromium produce a peak at 720 (LO) (longitudinal optical mode) cm^{-1}. With increasing thickness bands at 595 (LO), 412 (LO) and 610 (TO) (transverse optical mode) cm^{-1} appear. The main band at 720 cm^{-1} shifts to 760 cm^{-1} on increasing the thickness of the film. As deduced from TG and RBS data, Tables 2 and 3, the estimated thickness of the oxide

Figure 12 *Optical microscopy pictures of a sample after the following treatment: 2 minutes at 900°C; implantation of 10^{16} at/cm^2 of 100 keV La$^+$; 1 hour at 900°C. (A) and (B) indicate the spots at which the EDX was taken for Figure 13*

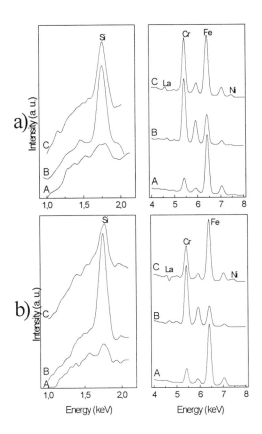

Figure 13 *EDX spectra of the following samples: (a) Preoxidized 2 minutes at 900°C; implantation of 10^{16} at/cm² of 100 keV La⁺; 1 hour at 900°C; (b) Non-preoxidized; implantation of 10^{16} at/cm² of 100 keV La⁺. The X-rays are taken in (A) the nodules, (B) the oxidized surface, (C) AISI304 as implanted (before oxidation)*

scale after preoxidation is over 100 nm. So, we can assign peaks at 416, 586 and 774 cm⁻¹ to Cr_2O_3. However, since α-Fe_2O_3 shows peaks at 613, 500 and 412 cm⁻¹ [13] partial substitution of chromium ions by iron cannot be disregarded, in any case a chromium-rich phase may be assumed since according to Music et al.[11] a shift to higher wave numbers has to be expected on increasing the chromium content (774 cm⁻¹) and a chromium-poor phase may be associated with the 730 cm⁻¹ peak. In addition to the chromia phase, chromium–manganese spinel phase is detected through the peak at 667 cm⁻¹ [10].

The IR spectrum of the stainless steel La-implanted sample (Figure 10, a) behaves quite similarly to that of the preoxidized specimen indicating that during the implantation process, oxidation of the surface layers occurs. The main difference is the higher intensity of the peaks around 700 and 482 cm⁻¹ that can be associated to the presence of chromite. If the implantation process is carried out over the preoxidized sample, Figure 10c, the elimination of the peak at

774 cm^{-1} is evident and the main peak appears now at 730 cm^{-1} indicating that chromium is partially removed from the surface layers during the implantation process, probably due to sputtering of the oxide layer.

After 1 hour oxidation at 900°C (Figures 10b and d) the two implantation procedures result in similar IR spectra. The only difference lays in a higher development of the chromite phase when lanthanum is implanted in the untreated stainless steel as deduced from the high intensity of the 645–673 cm^{-1} region.

In addition to these facts, either the implantation or the oxidation process results in the appearance of a set of bands in the 1200–900 cm^{-1} region that have to be ascribed to the formation of chromate-like phases[12]. These chromate-like phases have been previously detected on reacting metallic oxides present on the oxide scale with lanthanum nitrate[14].

An investigation of the surface of the samples with the optical microscope shows some interesting effects (Figures 11 and 12). First of all we can see that the granular structure, visible after the preoxidation treatment on the un-implanted sample (top of Figure 11), is swept away during the implantation (bottom of Figure 11), due to the erosion or sputtering provoked in the process. This microstructural change of the surface occurs in either preoxidized and non-preoxidized samples. Furthermore, all the specimens present oxide nodules after 60 minutes oxidation (Figure 12), and this fact can explain the shift of the Cr_2O_3 IR band. Although they have not been thoroughly studied, they could have been produced by the iron enrichment due to the chromium removal and/or the presence of secondary phases precipitated as a result of the implantation. The EDX analyses depicted in Figure 13 show that those granules are rich in iron, i.e. iron oxides (A), if we compare the composition with that of the rest of the surface (B), mainly chromium oxides. The segregation of silicon is another important feature of the oxidation process.

4 CONCLUSIONS

In this work, we have been able to prove the enhancement of the resistance to high temperature corrosion of AISI-304 stainless steel at the initial stages of the oxidation, when its surface composition is modified by means of the implantation of lanthanum at the proper doses. The implantation process alone also provokes the oxidation of the surface, as found by FT–IR, its erosion or sputtering, according to FT–IR and EDX and optical microscopy, and the modification of the surface structure seen by SEM and XRD.

During the oxidation process, the presence of iron oxide-rich nodules found by EDX and IR, may result, for longer periods of time, in voids that would lead to the spallation of the oxide scale. So, this study should be completed by analysing the next stages of the oxidation process, when the effect of the lanthanum would be more evident, and also after a post-implantation surface annealing for reducing existing damages.

Acknowledgements

Thanks are due to Luis Redondo and Jorge Rocha from the Instituto Tecnologico e Nuclear for providing the implanted samples. Financial support was provided by the European Carbon and Steel Commission (ECSC) under contract No. 7210/MA/948 and the Comisión Inter-ministerial de Ciencia y Tecnología (MAT95-1093-CE).

References

1. L. B. Pfeil, U.K. Pat. 574,088, 1947.
2. P. Y. Hou and J. Stringer, *J. Electrochem. Soc.*, 1987, **134**, 7.
3. M. J. Bennet, *J. Vac. Sci. Technol.*, 1984, **B2**, 4.
4. G. Bonnet, J. P. Larpin and J. C. Colson, *Solid St. Ionics*, 1992, **51**, 11.
5. M. I. Ruiz, J. Almagro, A. Heredia, J. botella, J. J. Benitez and J. A. Odriozola, in Processes & Materials Innovation. Stainless Steel. Associazione Italiana di Metalurgia, Milano (Italy) Vol. 3, 77, 1993.
6. T. Laursen, L. Claphan, J. L. Witthon and J. A. Jackman, Nucl. Inst. & Meth., 1991, **B59/60**, 768.
7. L. R. Doolitle, Nucl. Instr. Meth., 1985, **B9**, 344.
8. F. J. Ager, M. A. Respaldiza, J. Botella, J. C. Soares, M. F. da Silva, J. J. Benitez, J. A. Odriozola, *Acta Metall. Mater.*, 1996, **44**, 675.
9. D.K. Ottesen, *J. Electrochem. Soc.*, 1985, **132**, 2250.
10. M. Lenglet, R. Guillamet, J. Lopitaux and B. Hannoyer, *Mater. Res. Bull.*, 1990, **25**, 575.
11. S. Music, S. Popovic and M. Ristic, *J. Mater. Sci.*, 1993, **28**, 632.
12. M. Le Calvar and M. Lenglet, *Stud. Surf. Sci. Catal.*, 1989, **48**, 575.
13. D. Thierry, D. Persson, C. Leygraf, D. Delichere, S. Joiret, C. Pallotta and A. Hugot-Le Goff, J. Electrochem. Soc., 1988, **135**, 305.
14. M. I. Ruiz, A. Heredia, J. Botella and J. A. Odriozola, *J. Mater. Sci.*, 1995, **30**, 5146.

1.1.2

The Oxidation and Hot Corrosion Properties of Rare-Earth Modified Cr/Al Coatings on Fe- and Ni- Base Alloys

J. Kipkemoi and D. Tsipas

ARISTOTLES UNIVERSITY OF THESSALONIKI, DEPARTMENT OF MECHANICAL ENGINEERING, LABORATORY OF PHYSICAL METALLURGY, THESSALONIKI, 54006, GREECE

1 INTRODUCTION

High temperature oxidation and corrosion, create, for energy, aerospace, chemical and other industries, a huge annual direct cost, especially in maintenance and in unscheduled shutdowns. Surface modification[1] of materials used in these industries offers the best prospect of enhancing their resistance to high temperature degradation, without compromising their high temperature mechanical properties.

Ferritic and austenitic Fe-base alloys are choice materials in coal utilisation and energy conversion systems whereas Ni-base alloys find wide application in gas turbines. Design criteria and high cost limit the Cr and Al contents of these alloys to such low levels that it becomes necessary to apply coatings enriched with beneficial elements such as aluminium and chromium to increase their service life[2] . In less aggressive environments MX (where M = Ni, Fe and X= Al, Cr) coatings offer adequate protection. However the need to raise fuel efficiency, and the incentive to use low grade fuels[3], lead to more aggressive operating environments with higher temperatures and higher pS_2. In such environments the interaction of Cr and Al in MCrAl-type coatings provides a more effective protection, if the desired composition of Cr and Al is achieved. The Cr_2O_3 scale which forms initially on the metal's surface upon exposure, retards the internal oxidation of aluminium, whose atoms will, with time, reach the Cr_2O_3 scale/alloy interface to form a more protective, slow growing Al_2O_3[4].

Typical overlay-type (M-Cr-Al-Y) coatings usually contain 10–15 at.% Cr, 20–30 at.% Al and < 0.1 at.% Y[5] and hence provide superior protection. Diffusion coatings using the pack cementation technique do not have the line-of-sight restrictions and are 7–10 times cheaper[6]. In addition, they enjoy other distinct advantages including: superior adhesion between the coating and the substrate, applicability to components with complex geometries for example turbine blades with cooling channels. Traditional commercial practice involves pack aluminising, pack chromising and pack chromising-aluminising in a two step process. But the application of a Cr–Al coating in a one step process is attractive both economically and practically.

Many studies[4,6-11] have been carried out to codeposit beneficial elements Cr, Al, Si, by using this technique on the surfaces of both Ni-base and Fe-base alloys. Standard pack cementation components are; the substrate, a master alloy, an activator and an inert filler. The process involves four interrelated mechanistic steps[12]: (a) the thermodynamic equilibrium between the activator and master alloy, (b) the gaseous diffusion of metal halides from the pack to the substrate surface, (c) surface reactions at the substrate to deposit the coating elements and, (d) solid-state diffusion of the coating elements into the substrate. Coating growth is either inward or outward. In the case of outward growth the workpiece may be physically isolated from the pack to prevent the undesirable inclusion of pack powders.

Active elements – Table 1– are those which are added in small amounts to improve the oxidation resistance of metals and coatings especially under thermal cycling at high temperatures. They are characterised by their strong affinities for elements of the IVB and VIB groups. The beneficial effects of these elements – *reactive element effect* (REE) – have been well documented, and since their discovery, several mechanisms to explain this effect have been suggested and reviewed[13-16]. Two of these: (i) segregation of the reactive element to the scale grain boundaries and (ii) action of the reactive element in eliminating sulphur segregation, enjoy wider acceptance because of experimental supporting evidence. Overlay (M-Cr-Al-Y) type coatings have always included a reactive element. However the line-of-sight factor and high cost restrict them from broader applications[6,17].

Adverse effects of bulk additions of these elements are well known[3], therefore it is desirable to introduce them into a surface coating. Reactive elements can be introduced into a pack: (1) as an oxide (e.g. HfO_2, Y_2O_3) or (2) as an halide activator source (e.g. $HfCl_4$, YCl_3). The incorporation of Hf into the surface coatings using the pack cementation technique, on Fe-base alloys has been reported[19]. In early studies[8,17] and in more recent ones[18] only trace amounts of Y have been detected.

In the present study we preport the results of the high temperature oxidation and hot corrosion of RE-modified Cr/Al coatings produced by the pack cementation technique. The microstructures and morphologies obtained are emphasized.

Table 1 *Periodic table indicating those elements which have been found to act as "active elements"[21].*

IA	IIA	IIIA	IVA	VA	VIA	VIIA	⟵ VIII ⟶		IB	IIB	IIIB	IVB	VB	VIB	VIIB	O
H																He
Li	Be										B	C	N	O	F	Ne
Na	Mg										Al	Si	P	S	Cl	Ar
K	Ca	Sc	Ti	V	Cr	Mn	Fe	Co	Ni	Cu	Zn	Ga	Ge	As	Se	Br	Kr
Rb	Sr	Y	Zr	Nb	Mo	Tc	Ru	Rh	Pd	Ag	Cd	In	Sn	Sb	Te	I	Xe
Cs	Ba	La	Hf	Ta	W	Re	Os	Ir	Pt	Au	Hg	Tl	Pb	Bi	Po	At	Rn
Fr	Ra	Ac															
	Ce	Pr	Nd	Pm	Sm	Eu	Gd	Tb	Dy	Ho	Er	Tm	Yb	Lu			
	Th	Pa	U	Np	Pu	Am	Cm	Bk	Cf	Es	Fm	Md	No	Lr			

2 EXPERIMENTAL PROCEDURE

2.1 Materials

The alloys selected for this study were two Ni-base alloys namely RENE 80 and INC718. The first is a high strength structural alloy with application mainly in gas turbines as material for blades and vanes. The second has a lower nickel content and a substantial amount of iron, which limits its application to lower temperatures. A low alloy steel, 2.25 Cr-1Mo which is widely used as boiler tube material was also chosen. Samples with dimensions of about 1.5 x 1.0 x 0.5 cm were machined from each alloy in the as-received condition, and given standard preparation for the coating process.

2.2 Pack Cementation Process

The experimental procedure for aluminising, and codeposition of Cr and Al with and without Hf modification has been described elsewhere[19,20]. Table 2 gives the pack powder compositions used in the present study, where we attempted to introduce Y, Cr and Al, into the surface of 2.25 Cr-1Mo steel. Alloy samples, were placed in the appropriate retorts and kept at process temperature of 1150°C for 8 hours.

Gaseous species in halide-activated cementation packs play a decisive rôle in the deposition process. The requirement for a binary masteralloy powder if both elements are to be codeposited into a substrate simultaneously is that each element should have significant and comparable partial pressures[11]. This is achieved by using a suitable activator, and properly controlled thermodynamic activities of the components of the masteralloy. The masteralloy in this study was 92.5Cr–7.5 Al, and the activator was a mixture of different proportions of NH_4Cl and YCl_3. The yttrium salt was also the reactive element source. In one selected pack Y metal powder was added.

2.3 Oxidation /Hot Corrosion Testing

To evaluate and study the oxidation and hot corrosion properties of both the coated and the uncoated alloy samples, testing and characterisation were done as described in an earlier report[18]. In the present report, similar testing conditions were adopted for comparison purposes. Specifically, cyclic oxidation was conducted in air. The furnace temperature was 800°C. The cycle control device was set such that the hot cycle was 30 minutes and the cool cycle was 15 minutes. The kinetic behaviour was presented in terms of weight changes versus the number of cycles.

Table 2 *Pack powder composition for coating process on 2.25 Cr-Mo steel*

Packs	Composition (wt. %)				
	92.5Cr 7.5Al	NH_4Cl	YCl_3	Y	Al_2O_3
1	15	2	2	-	Bal
2	15	1	3	-	Bal
3	15	0.6	3.4	-	Bal
4	15	2	2	4	Bal

To induce hot corrosion, alloy samples were preheated to about 200°C, and air sprayed with a saturated aqueous solution of Na_2SO_4. Evaporation of the solvent left a thin salt coating on the surfaces of the samples which were then cycled. The melting point of Na_2SO_4 is 884°C, therefore the furnace temperature of 875°C chosen, ensured that the salt was just molten. Periodically it was necessary to replenish the salt coating. For comparison purposes some samples were not coated with the salt.

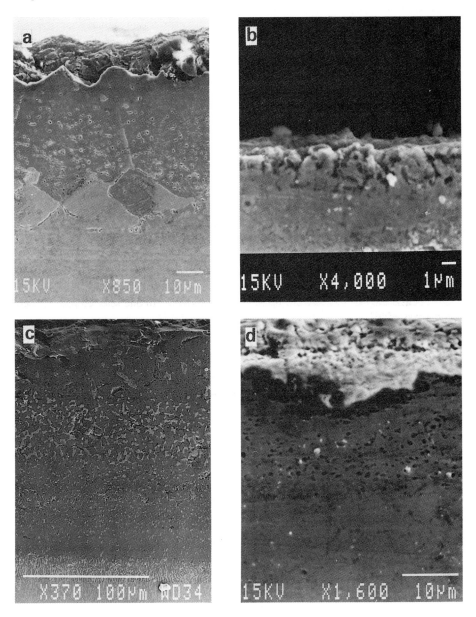

Figure 1 *SEM morphologies after hot corrosion at 870°C*

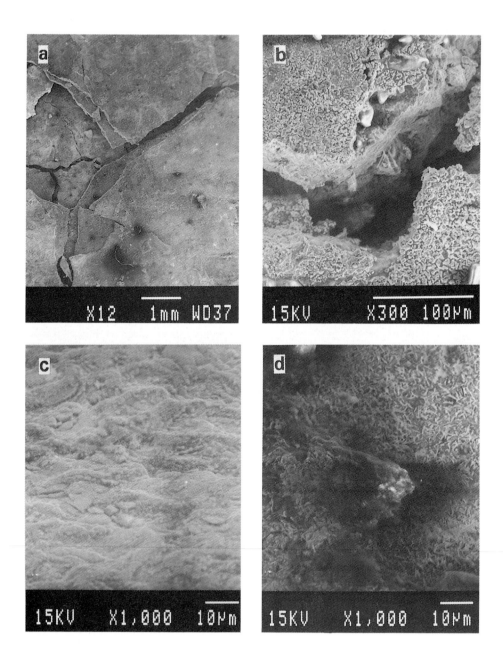

Figure 2 *SEM surface morphologies of 2.25 Cr-Mo steel, (a) uncoated – after cyclic oxidation at 875ºC, (b) uncoatded – after hot corrosion at 875ºC, (c) Cr-Al-Hf coating after cyclic oxidation at 800ºC and (d) Cr–Al coating after cyclic oxidation at 800ºC*

3 RESULTS AND DISCUSSION

3.1 Diffusion Aluminides

Figure 1 shows the microstructures and morphologies of: (a) INC 718 (aluminised), (b) INC 718 (uncoated), (c) RENE 80 (aluminised) and (d) RENE 80 (uncoated) after the Na_2SO_4 induced hot corrosion test. The aluminide coatings had a 3-zone structure. The inner zone consisted of an aluminium-rich NiAl matrix and carbide precipitates. The outer zone consisted of an aluminium-rich βNiAl matrix dispersed with substrate element precipitates. Surface morphology of the coating on INC 718 indicated a substantial material removal, however the integrity of the coating below the corroded surface remained intact with no signs of internal oxidation, sulphur penetration or cracks. The coating on RENE 80 had minimal material removal. Lateral voids were observed, probably arising from Al depletion and internal microcracking. However the coatings maintained (a) sufficient ductility and compatible thermal expansion with the substrate to prevent extensive cracking and (b) low interdiffusion rates

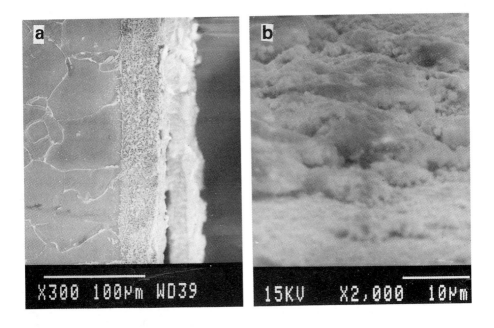

Figure 3 *SEM morphologies of 2.25Cr-1Mo steel with a Cr-Al coating after cyclic oxidation at 800°C: (a) cross-section, (b) surface*

between coating and substrate and minimum cation transport to inhibit coating deformation through phase transformation. The oxide scale formed on the uncoated INC 718 was thin and spalled easily. This alloy contains about 18 wt.% Fe and forms upon exposure a dusty non-adherent oxide. Sulphur was detected up to 5μm below the surface. Corrosion products embedded into the oxide scale remained on the surface of the uncoated RENE 80, to give the morphology which is characteristic of the salt fluxing degradation mode[20].

3.2 RE-modified Cr-Al coatings

Low alloy steels are austenitic at coating temperature of 1150°C, and the incorporation of Cr and Al which are both ferrite-stabilisers results in the austenite-ferrite transformation[4]. So the region immediately below the coating/substrate interface is always decarburized as the carbon is rejected into the substrate's core.

Figures 2a and 2b show the morphology of the uncoated low alloy steel surfaces after cyclic oxidation and Na_2SO_4-induced cyclic hot corrosion respectively, at 875°C. The alloy suffered extensive damage. The morphology was that of several layers of Iron-rich oxide scales formed during successive cycles. They were held loosely together on the surface. The corrosion products which included the unconsummed Na_2SO_4 melt, recrystallised to form the white crystals seen on Figure 2b.

3.2.1 Hf-Modified Coatings. The microstructures and compositions of Cr-Al coatings successfully modified by Hf addition have been presented in an earlier paper[19]. Figures 2 and

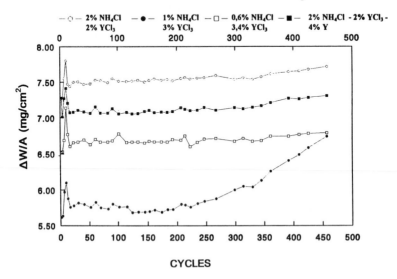

Cyclic Oxidation of Fe-base alloy at 800°C

Figure 4 *Specific weight changes for 800°C cyclic oxidation in air coated 2.25Cr-1Mo steel versus the number of cycles*

3 show the morphologies of the Cr-Al-Hf coating and (Figures 2d, 3a and 3b) of the Cr–Al coating on the low alloy steel after the cyclic oxidation test. The coatings suffered extensive deformation, however there were no cracks, neither on the protective oxide scale nor within the coating. When coatings are heated at elevated temperatures certain elements leave the coating to form oxides at the surface. Internal oxidation may also take place. These changes cause phase transformation which may lead to coating deformation and eventual breakdown. But this form of coating degradation is less severe than cracking and spalling. The Hf-free Cr-Al coating had cracks which extended from the oxide scale to the coating. Cracking in coatings exposed to thermal cycling is primarily associated with the compressive and tensile stresses generated during heating and cooling. The coating thickness and plasticity are also important factors in cracking. Thicker coatings and those with poor plasticity tend to crack and spall easily. Active elements are known to modify scale morphology and structure[16] leading to a texture with enhanced scale plasticity and growth stress relief. The surface morphology, Figure 2c, of the Hf-modified coating indicated that a mild-type mechanism of material removal, was operational compared to the Hf-free Cr-Al coating surface, Figures 2d and 3b.

3.2.2 Y-Modified Coatings. Figure 4 is a plot of the weight change behaviour of the coatings obtained from the four packs given on Table 2. To achieve the addition of a third element during a simultaneous codeposition of Cr–Al, two conditions must be met[11]: (i) the third element must have a reasonable solubility in the substrate to be coated. For Fe substrate, Y and Hf each exhibit solubility compatible with the desired range of composition and (ii) the third element should not coexist in the pack as a pure metallic phase – hence the metallic element is usually added in its oxide form, or as a halide salt.

To investigate the effects of adding a pure metallic phase, Y metal powder was added in one of the packs. Kinetic results at 800°C, however did not show any adverse effects. All coatings obtained had rather low Al content, due perhaps to the composition of the masteralloy used, compared to the masteralloy in other studies[19] which was more rich in Al. Furthermore, competition arising from the mutual presence of the different gaseous species in the pack resulted in the overall reduction of the individual partial pressures of the metallic halides, as pointed out by Kung et al[11].

Preliminary EPMA detected only trace amounts of yttrium. But the overall performance of these coatings was superior compared to the RE free Cr–Al coatings studied earlier[19] under the same cyclic oxidation conditions. The coatings from pack number 2, with the ratio $NH_4Cl:YCl_3$ = 1:3 produced the best overall performance.

4 CONCLUSIONS

Codeposition of Cr and Al on the surfaces of Ni-base and Fe-base alloys was conducted. It was possible to introduce the third element into the halide-activated pack as an oxide, or as a halide activator, enabling the deposition of three elements simultaneously in a one step pack cementation process. The diffusion coatings obtained had among other attributes, excellent coating/substrate adhesion.

Coating structure and morphology indicated that the active elements influenced the oxide scale growth and this may be responsible for the improvement of the oxidation and hot corrosion properties of the coatings.

References

1. E. D. Hondros in P. K. Datta and J.S. Gray (eds) 'Surface Engineering'. Vol.1. Royal Society of Chemistry, Cambridge. U.K., 1993, p. 1.
2. W. Da Costa, B.Gleeson and D. J. Young. *J. Electrochem. Soc.*, 1994, **141**, 1464.
3. A. D. Zervaki, C. N. Haidemenopoulos and D. N. Tsipas, in P.K. Datta and J.S. Gray (eds) 'Surface Engineering' Vol 1.Royal Society of Chemistry, Cambridge.U.K.,1993, p. 71.
4. R. Bianco, M. A. Harper, and R. A. Rapp. *JOM.* Nov. 1991, 68.
5. J. A. Nesbitt and R.W. Heckel, *Thin solid films*, 1984, **119**, 281.
6. B. Gleeson, W. H. Cheung, W. Da Costa and D. J. Young, *Oxidation of Metals,* 1992, **38**, 407.
7. E. Fitzer and H. J. Maurer, in 'Materials to Resist High Temperature Corrosion', Applied Science London, 1978, p. 253.
8. R. A. Rapp, D. Wang and T. Weisert. in M. Khobaib and R. Krutenat (eds) 'Metallurgical Coatings', TMS–AIME, Warendate PA., 1987, p. 131.
9. D. M. Miller, S. C. Kung, S. D. Scarberry and R. A. Rapp. *Oxidation of Metals,* 1988, **29**, 239.
10. P. A. Choquet, M. A. Harper and R.A. Rapp, *Journal De Physique*, Suppl. No 5, Tom., 1989, 50, 5, C5-681
11. S. C. Kung and R. Rapp, *Oxidation of Metals*, 1989, **32**, 89.
12. S. C. Kung and R. Rapp, *J. Electrochem. Soc.*, 1988, **135**, 731.
13. B. Pieraggi and R. Rapp, *J. Electrochem. Soc.*, 1993, **140**, 2844.
14. J. B. Pfeil. UK. Pat. 459, 848, 1937.
15. Y. Saito and B. Onay, *Surf. and Coatings Technology*, 1990, **43/44**, 336.
16. D. P. Whittle and J. Stringer, *Phil. Trans. R. Soc. Lond.*, 1980, **A295**, 309.
17. D. C. Tu, C. C. Jin, S. J. Liao and J. C. Chou, *J. Vac. Sci. Technol.*, 1986, **A4**, 2601.
18. R. Bianco and R.A. Rapp, *J. Electrochem. Soc.*, 1993, **140**, 1184.
19. J. Kipkemoi and D. Tsipas (in print).
20. H. L.Du, J. Kipkemoi and D. Tsipas and P. Datta, Paper presented at the International Conference on Metallurgical coatings, San Diego. CA. USA, April, 22–27. 1996.
21. G. Beranger, F. Armet and M. Lambertin in E. Lang (ed) 'The Role of Active Elements in the Oxidation Behaviour of High Temperature Metals and Alloys'. Elsevier Applied Science, 1989, p. 34.

1.1.3
The Effect of H_2O, O_2 in Atmosphere and Cr, Si in Steel on the Oxidation Behavior of Ferritic Stainless Steels

Hisao Fujikawa[1] and Yoshiaki Shida[2]

[1]STEEL SHEET AND PLATE DIVISION
[2]ADVANCED TECHNOLOGY RESEARCH LABORATORIES. SUMITOMO METAL INDUSTRIES LTD.. FUSOH-CHO 1–8, AMAGASAKI, 660, JAPAN

1 INTRODUCTION

Ferritic stainless steels have good oxidation resistance up to a considerably high temperature. For example, 18%Cr steel shows good oxidation resistance up to 800°C[1,2]. However, it is well known that the oxide scale shows a two-layer structure – Fe_3O_4 in the outer layer and $(Fe,Cr)_3O_4$ in the inner layer. This scale structure is generally observed in austenitic stainless steels as well as in ferritic ones[3-5]. Furthermore, the oxidation rate is more rapid in water vapor than in the O_2 atmosphere.

Less than 10%O_2 and 10-20%H_2O exist in the combustion environment. The oxidation behavior of stainless steels in the combustion environment in which O_2 and H_2O exist is an important problem. For example, when type 430 steel is used in the kerosene wick of combusted heaters, this steel produces the reddish brown oxide; this lowers the efficiency of the fuel combustion rate.

The oxidation behavior of ferritic stainless steels in H_2O and O_2 mixed atmosphere has been studied by Kawasaki et al. and by the authors [6-10]. It has already been shown that the observed reddish brown scale is composed of nodular Fe_2O_3, Fe_3O_4 and $(Fe,Cr)_3O_4$ layer structure from the gas side, and then grows all over the surface. This extreme behavior can be observed at about 600°C in low O_2 pressure and in high H_2O pressure atmosphere. The authors have also shown that Cr and Si in steels are an effective way of preventing the reddish brown oxide scale. In H_2O and O_2 mixed gas atmosphere, H_2O promotes the growth of nodular Fe_2O_3 under the existence of O_2. However, the synergism of H_2O and O_2 on the formation of initial oxide scale and the effect of alloying elements on the oxidation behavior of stainless steels have not been exactly clear.

First, in the present study, the effect of H_2O and O_2 in the atmosphere on the oxidation behavior of ferritic stainless steels was examined. Next, the relationship between the composition of the initial oxide scale and the nodular oxide scale observed in the abnormal oxidation was studied. Finally, the effect of Cr and Si in steels on the scale composition was also studied.

Table 1 *Chemical composition of test materials (wt%)*

No.	C	Si	Mn	P	S	Cr	Nb
1	0.010	0.58	0.48	0.023	0.001	16.97	-
2	0.010	2.10	0.52	0.024	0.001	11.09	-
3	0.016	2.58	0.24	0.020	0.006	18.23	0.38
4	0.009	2.22	0.34	0.014	0.002	03.60	0.30
5	0.001	0.30	0.30	0.003	0.007	14.65	-
6	0.005	1.97	0.30	0.003	0.007	14.74	-
7	0.001	3.00	0.30	0.003	0.007	14.75	-
8	0.011	1.57	0.25	0.024	0.002	16.41	0.29
9	0.011	2.06	0.25	0.022	0.002	16.37	0.29
10	0.011	2.59	0.25	0.018	0.002	16.33	0.28
11	0.012	2.76	0.25	0.019	0.002	16.35	0.28
12	0.014	3.06	0.25	0.026	0.002	16.48	0.29
13	0.004	0.34	0.30	0.003	0.007	17.86	-
14	0.011	1.56	0.25	0.024	0.002	18.03	0.30
15	0.015	2.08	0.25	0.024	0.001	18.04	0.28
16	0.015	2.57	0.25	0.020	0.002	18.03	0.28
17	0.014	2.74	0.25	0.021	0.002	18.12	0.27
18	0.012	3.06	0.25	0.021	0.002	18.24	0.28
19	0.013	0.46	0.27	0.021	0.002	20.00	0.22
20	0.014	0.96	0.27	0.023	0.002	20.59	0.24
21	0.014	1.58	0.27	0.023	0.002	20.53	0.24
22	0.014	2.08	0.26	0.022	0.002	20.34	0.24
23	0.010	2.62	0.26	0.022	0.002	20.19	0.24
24	0.013	0.95	0.27	0.023	0.002	22.50	0.25
25	0.011	1.55	0.27	0.023	0.002	22.30	0.24
26	0.013	2.02	0.26	0.022	0.002	22.27	0.23
27	0.015	2.58	0.26	0.023	0.002	22.30	0.23

2 PREPARATION OF TEST SPECIMENS AND EXPERIMENTAL METHODS

2.1 Preparation of Test Specimens

25 steels containing 11~22.5%Cr and less than 3%Si as shown in Table 1 were used. These steels mainly contain low C and a little Nb. No.1 to 3 steels were conventionally produced in cold rolled sheets of 0.8 mm in thickness. No.4 to 25 steels were melted in a vacuum induction furnace in our laboratories. They were forged, hot rolled, heat treated, then cold rolled in 0.8 mm in thickness, and were also heat treated. After that, the sheets were pickled (in 3%HF + 10%HNO$_3$ solution). The sheets were cut in 10 mm by 20 mm. The surface of the test specimens was polished with No.1,200 emery paper. Lastly, the test specimens were also pickled in order to remove the mechanical finishing layer.

2.2 Experimental Method

The oxidation test was carried out using the equipment shown in Figure1. The test specimens were hung at the center of the quartz reactor. Ar + X%O$_2$ gas was the standard gas (X being 2,5,10,15 and 20) . The gas was streamed into the water in a constant temperature vessel. It

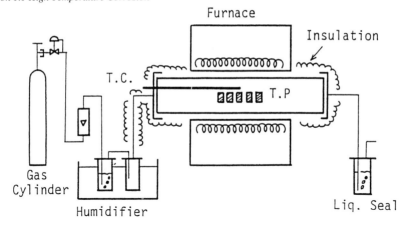

Figure 1 *Schematic diagram for the test apparatus*

was humidified and was then streamed into the reactor furnace. The pipe from the humidifier to the reactor tube was covered with a ribbon-type heater and was kept at above 100°C in order to prevent condensation in the pipe.

Ar + X%O$_2$ mixed gas was used as standard gas. Y%H$_2$O (Y means 0, 10, 15 and 20) was introduced into this gas which was in the humidifier. Gas velocity was kept at 200 ml/min.

After the oxidation test, the weight change of the test specimens was recorded and the formation of the reddish brown oxide scale was examined. The study of the oxidation behavior of steels was carried out precisely, using SEM (scanning electron microscope), optical microscope, TEM (transmission electron microscope), STEM (scanning transmission electron microscope) and IMMA (ion mass microanalyzer).

3 EXPERIMENTAL RESULTS

3.1 The Effect of O$_2$ and H$_2$O Concentration in the Atmosphere

3.1.1 The effect of O$_2$. The oxidation behavior of steels was examined in the atmosphere in which O$_2$ concentration was changed from 0 to 20% in Ar + 20%H$_2$O standard gas. As shown

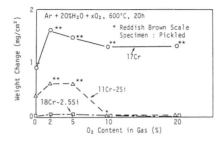

Figure 2 *Effect of O$_2$ on the oxidation behavior of ferritic steels in 20%H$_2$O containing atmosphere*

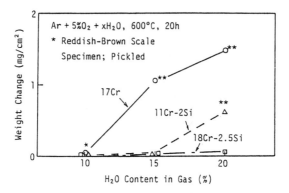

Figure 3 *Effect of H_2O on the oxidation behavior of three ferritic steels in 5%O_2 containing atmosphere*

in Figure 2, the 17%Cr steel showed a larger weight gain and produced more reddish brown oxide scale in the atmosphere containing O_2 than in the one containing 0%O_2. Remarkably in the case of the O_2 containing atmosphere, it was shown that the weight gains were almost the same, in spite of the O_2 concentration. The weight gain of 11%Cr-2%Si steel also increased more in a (2–5%)O_2 than in 0%O_2. In this steel, the weight gain decreased with the increase of O_2 and the reddish brown oxide scale was observed in 2–5%O_2. This steel had less weight gain and showed the reddish brown oxide scale in narrower O_2 range than 430 type steel (17%Cr steel). While, 18%Cr-2.5%Si steel showed good oxidation resistance and did not show the reddish brown oxide scale in all of the test conditions.

In the above results, the increase of weight gain and the degree of the formation of the reddish brown oxide scale were coincidental phenomena. A nodular reddish brown oxide scale was observed as will be shown later. That is, to conclude that when nodular oxide is formed, the oxidation resistance of steels is lowered.

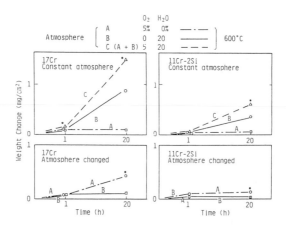

Figure 4 *Effect of separate exposure in O_2-containing and H_2O containing atmosphere on the oxidation behavior*

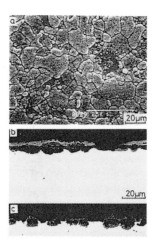

Figure 5 *Typical microscopic appearance of nodular scale formed on 17%Cr steel at 600 after 20 hrs, (a) Surface, 5%O_2 + 20%H_2O atmosphere, (b) Cross section, 5%O_2 + 20%H_2O atmosphere and (c) Cross section, 20%H_2O atmosphere*

3.1.2 The effect of H_2O. As mentioned above, the oxidation resistance of steels is lowered by the existence of a little O_2 in the atmosphere. Next, the effect of H_2O was examined by decreasing the O_2 concentration from 20% to 10% in Ar-5%O_2 standard gas. As shown in Figure 3, the 17%Cr steel showed a decrease in weight gain with the decrease of H_2O concentration and remarkably showed the harmful effect of H_2O. However, the reddish brown oxide scale was observed in more than 10% H_2O. 11%Cr-2%Si steel did not form the reddish brown oxide scale in less than 15%H_2O. 18%Cr-2.5%Si steel did not show the reddish brown oxide scale in any of the test conditions. Therefore, it was demonstrated that 18%Cr-2.5%Si steel has a high resistance to the formation of the reddish brown oxide scale in environments containing O_2 and H_2O.

3.1.3 The synergism of O_2 and H_2O. The next examination was carried out to clarify the effects of the synergism of O_2 and H_2O on the formation of the reddish oxide scale clear. That is, effects of O_2 and H_2O were separately examined. The test specimens were exposed in H_2O containing (0%O_2) Ar gas after being exposed in O_2 containing (0%H_2O) Ar gas, or were exposed in reverse gas. In these cases, it was studied how the oxide scale changed. 17%Cr (430 type) and 11%Cr-2%Si steels were used in these tests. As shown in Figure 4, the weight gain increased in the order of 5%O_2 (A), 20%H_2O (B)and 5%O_2 + 20%H_2O (C) containing Ar gases. It became clear that the weight gain increases in O_2 and H_2 mixed atmosphere.

Next, the atmosphere was changed in the early (0–1 hr) and later (1–20 hrs) periods of the oxidation tests. As shown in Figure 4, the reddish brown oxide scale was formed on the test specimens which were exposed to B (20%H_2O) gas in the early period and to A (5%O_2) gas in later periods. However, the weight gain was less than when they were exposed to C gas. The weight gain did not significantly increase in the case in which the test samples were exposed to A gas in the early period and to B gas in the later period. That is, it was found that the initial oxide scale in O_2 atmosphere was very stable.

From the above results, it could be concluded that the simultaneous attack of O_2 and H_2O accelerates the oxidation rate of steels.

3.2 Oxidation Mechanism

3.2.1 Occurrence and growth behavior of the reddish brown oxide scale. The surface and cross sections of the reddish brown oxide scale formed on 17%Cr steel are shown in Figure 5. The nodularly formed oxide scale covered over the grains of the steel gradually. However, the growth of the oxide scale was prevented by the grain boundary. This is due to the Cr rich oxide scale being initially formed at the grain boundary, as shown later. The cross sections of the cases of 20% H_2O and 5%O_2 + 20%H_2O atmospheres are shown. The weight gain in the case of the 5%O_2 + 20%H_2O was larger than that in the case of the 20%H_2O. The nodular two layer oxide scale was locally observed in the case of the 20%H_2O. On the other hand, the two layer oxide scale which grew laterally was observed in the case of the 5%O_2 + 20%H_2O. It could be suggested that the buds of the nodular oxide increase and rapidly grow laterally in the case of the O_2 and H_2O mixed atmosphere, as compared with the case of the O_2 atmosphere.

There are theories about the origin of the nodular oxide scale. For example, one such theory suggests that when the oxide scale cracks and the Cr depleted metal surface is exposed to the atmosphere, the Fe rich oxide scale is nodularly formed at the small fissures. Consequently, this point was examined next. This was confirmed in the three steels. After the test specimens of these steels were oxidized in a 5%O_2 + 20%H_2O atmosphere for one hour, two kinds of treatments were done on those specimens. In one treatment the specimens were descaled. In another treatment the oxide scale was cracked by bending the specimens and then the metal surface was exposed at the fissures. After these treatments, the specimens were exposed to same atmosphere for one hour and the occurrence of the reddish brown oxide scale was examined. As shown in Figure 6, in both 17%Cr and 11%Cr-2%Si steels, the occurrence of the reddish brown oxide scale was accelerated when using both treatments; particularly the remarkable nodular oxides were observed at the fissures. From this result, it could be predicted that when a defect such as a crack occurs in the scale, the reddish brown oxides easily form at this part. However, the reddish oxides were not observed in 18%Cr-2.5%Si steel in either treatments. Therefore, the reddish brown oxides do not occur in the high resistant steel, even if the defects exist in the oxide scale.

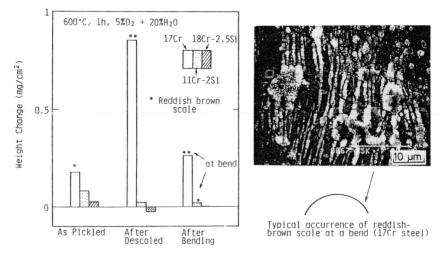

Figure 6 *Oxidation behavior of specimens after descaling or bending the initially formed scale (after 1 hr)*

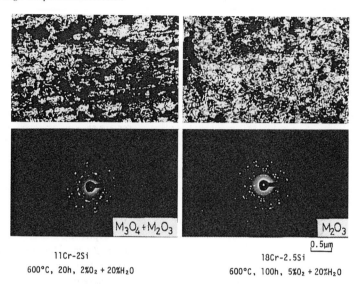

Figure 7 *Examples of transmission electron micrograph for thin oxide scale formed on Si containing ferritic steels in H_2O-O_2 containing atmosphere*

3.2.2 Analysis of Thin Oxide Scale Before the Reddish Brown Oxides Formed

The normal oxide part also locally existed in the test specimens in which the reddish brown oxides occurred. The normal oxide scale was very thin at this part. Therefore, analysis of this thin oxide scale is very important.

The thin oxide scale was extracted by the replica method and was observed with the electron microscope, as shown in Figure 7. The oxide scale of 11%Cr-2%Si steel formed in a 2%O_2 + 20%H_2O atmosphere was composed of a M_2O_3(Cr_2O_3) type and a M_3O_4(Fe-Cr spinel) type oxide. The oxide scale of 18%Cr-2.5%Si steel was composed only of the M_2O_3 type oxide. M_2O_3 is Cr_2O_3 dissolving a little Fe and M_3O_4 is a Fe-Cr spinel. The oxide scale formed on

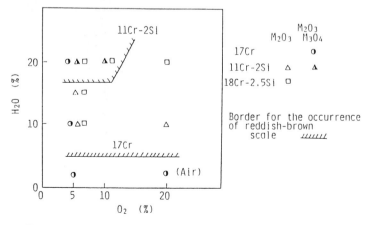

Figure 8 *Composition of thin scale vs. O_2/H_2O condition relationship for oxide scale found in H_2O-O_2 containing atmosphere*

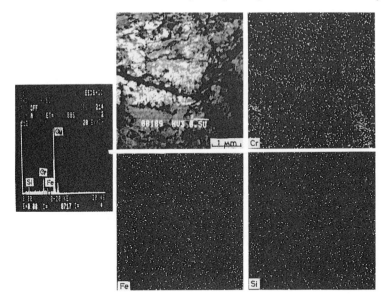

Figure 9 *Analytical transmission electron microscope image of thin scale formed on 11%Cr - 2%Si steel in 2%O_2 + 20%H_2O atmosphere at 600 for 20 hrs*

17%Cr, 11%Cr-2%Si and 18%Cr-2,5%Si steels in various kinds of atmospheric conditions are shown in Figure 8. Both M_2O_3 and M_3O_4 were detected on 17%Cr steel which was exposed in all kinds of atmospheric conditions and on 11%Cr-2%Si steel which was exposed to a little O_2 and a greater quantity of H_2O bearing atmosphere. The oxide scale formed of 11%Cr-2%Si steel was composed only of M_2O_3 in a greater quantity of O_2 atmosphere in spite of the H_2O content. Therefore, it could be suggested that O_2 accelerates the formation of M_2O_3 and H_2O accelerates the formation of M_3O_4. The occurrence of the region of the reddish brown oxide scale in 11%Cr-2%Si steel was shown in this figure. It is found that the formation of the region of M_3O_4 coincides with the occurrence region of the reddish brown oxide scale. Furthermore, in 18%Cr-2.5%Si steel, only M_2O_3 was detected in all atmospheric conditions.

The distribution of elements in the thin oxide scale, formed on 11%Cr-2%Si steel, was detected with the scanning transmission electron microanalyser (STEM). As shown in Figure 9, the oxide scale was composed of Cr, Si, Fe and O. Cr was enriched at the grain boundaries. (Cu shown in EDAX was introduced from the mesh.) It could suggest that this Cr enrichment prevents the development of the nodular oxide scale at the grain boundaries.

3.3 Effect of Cr and Si Contents in Steels

3.3.1 Effect of Cr and Si contents on the oxidation resistance of steels. The effect of O_2 and H_2O in the atmosphere and the beneficial effect of Cr and Si contents have already been mentioned above. Subsequently, the effect of Cr and Si was examined in a little O_2 and a greater quantity of H_2O atmosphere, using the steels shown in Table 1, in which the reddish brown oxide scale occurs most severely. As shown in Figure 10, it is clear that the increase of Cr and Si contents enhances the oxidation resistance of steels. The oxidation resistance of 15 - 18%Cr steels was remarkably enhanced by adding 2–3%Si. The oxidation resistance of more than 20%Cr steels was improved by adding about 1%Si.

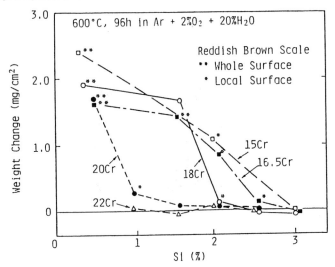

Figure 10 *Effect of Cr and Si content on the oxidation resistance of ferritic steels in a H_2O-O_2 containing atmosphere*

3.3.2 The effect of Cr and Si on the composition of initial thin oxide scale. The composition of the initial thin oxide scale was examined and it was examined to study how the thin oxide scale composition changes by applying Cr and Si in steels. The result of the electron diffraction analysis is shown in Figure 11. The region of M_3O_4 which was detected coincided with the region of the reddish brown oxide scale, as in Figure 8. Therefore, preventing the generation of M_3O_4 is related to the improvement of the oxidation resistance of steels.

The oxide scale composition was analyzed by IMMA. In the example of 18%Cr-2,5%Si steel in Figure 12, the concentrations of Cr , Si and Mn were detected in the outermost surface layer. The diluted Cr and enriched Fe were detected in the inner layer. Cr was remarkably enriched at the metal/oxide scale interface in the innermost layer. It was determined that the Cr enriched innermost layer prevents the oxidation of Fe. However, a Si enriched layer was

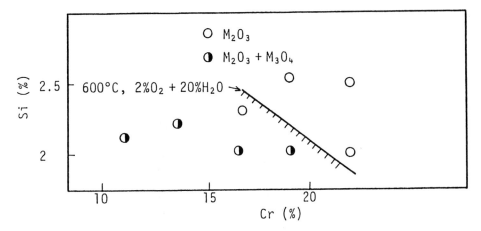

Figure 11 *Composition of thin scale vs. Cr and Si content relationship for oxide scale formed in H_2O-O_2 containing atmosphere*

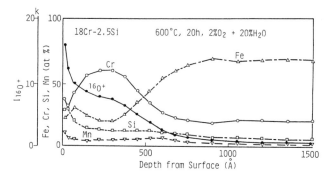

Figure 12 *An example of IMMA depth analysis profile of thin scale formed on 18%Cr - 2.5%Si steel*

not detected at the inner oxide scale layer/metal interface.

Cr and Fe concentrations at about 225 Å from the metal surface in the innermost scale were analyzed in the steels tested and then, the relationship between the Cr/Fe ratio in the innermost scale and the oxidation resistance of steels was examined, as shown in Figure 13a. In this figure, Si concentration at about 4 Å from the metal surface was also shown. The oxidation resistance of steels was enhanced by the increase of Cr and Si contents and also with the increase of Cr/Fe ratio. Si concentration in the outermost scale did not coincide with the oxidation resistance of steels. The increase of Cr/Fe ratio resulted from the increase of Si concentration. Therefore, it could be suggested that Si has the effect of promoting Cr enrichment.

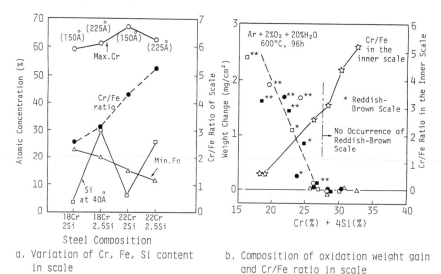

Figure 13 *Influence of scale composition on the oxidation resistance of ferritic steels in H_2O-O containing atmosphere, (a) Variation of Cr, Fe and Si content in scale (b) Composition of weight gain and Cr/Fe ratio in scale*

Next, the relationship between the occurrence of the reddish brown oxide scale and the Cr/Fe ratio was studied. Before this study, the equivalent relationship between Cr and Si contents for the oxidation resistance of the steels was calculated from Figure 10. It was found that the effect of Si for the prevention of the reddish brown oxide scale was about 4 times stronger than that of Cr. Therefore, Figure 13b was described by using Cr(%) + 4Si(%) as the parameter. WhenCr + 4Si ≥ 27, the occurrence of the reddish brown oxide scale was not observed. Also Cr/Fe ratio was above 3 in this case.

From these results, the M_2O_3 type oxide layer has to be formed homogeneously (or uniformly) in order to prevent the formation of the reddish brown oxide scale. Therefore, the Cr/Fe ratio in the Cr rich layer has to be above 3. Furthermore, Cr and Si contents in the steels have to be Cr+ 4Si ≥ 27.

4 DISCUSSION

4.1 The Effect of O_2 and H_2O on the Composition of the Initial Thin Oxide Scale

It is clear that O_2 and H_2O promote the formation of M_2O_3 and M_3O_4 type oxides. This result is significantly connected with the fact that the nodular oxide scale occurs and the steels show bad resistant to oxidation, when M_3O_4 type oxide occurs in the initial oxide scale.

One question is why M_3O_4 type oxide occurs in a greater quantity of H_2O atmosphere in the steels which produce M_2O_3 type oxide in a little H_2O atmosphere. It can be suggested that the reason is that the oxide scale is apt to yield defects such as cracking due to the attack by H_2O. It is thought that H_2 gas occurs in the oxide scale and the oxide scale becomes brittle when exposed to H_2 gas. It can be easily determined that when the defects occur in the oxide scale, the M_3O_4 type oxide occurs in those parts as shown in Figure 6. When the M_3O_4 type oxide is formed, the diffusion rate of metal ions in the M_3O_4 type oxide is much more rapid than that in the M_2O_3 type. For example, it is 10^4 to 10^5 faster at 600°C in the extrapolation from the high temperature. Therefore, it can be easily understood that M_3O_4 type oxide rapidly develops and leads to the nodular oxide scale.

4.2 The Effect of Cr and Si on the Improvement of the Oxidation Resistance of the Steels

Cr and Si are beneficial elements for the improvement of the oxidation resistance of the steels, as mentioned above. It was demonstrated that when the Cr/Fe ratio in the Cr rich layer in the oxide scale is 3, M_2O_3-type oxide scale is homogeneously formed and the formation of the nodular oxide scale cannot be observed. It is interesting that when Cr/Fe is 3, M_2O_3-type oxide was detected. $FeCr_2O_4$ which has the highest concentration of Cr content in the M_3O_4-type oxide is Cr/Fe = 2. It could be suggested that when Cr/Fe is 2, the oxide is either a mixture of Cr_2O_3 and $FeCr_2O_4$ or $(Cr,Fe)_2O_3$. However, it is found from this study that only the M_2O_3-type oxide is stable. Mortimer et al. also already showed that when Cr/Fe is 3, Cr_2O_3-type oxide is detected from the analysis of the oxide scale produced on Fe-20 - 38%Cr alloys heated at 650–950°C. Therefore, it could be suggested that the transformation from spinel-type oxide to Cr_2O_3-type oxide is thermodynamically stable at Cr/Fe = 3.

Since high Cr/Fe ratio in the scale improves the oxidation resistance of steel, high Cr bearing steel has excellent oxidation resistance in $H_2O + O_2$ mixed atmosphere. However, in that case, the Cr content in the steel needs to be 25%. Next, it is found that Si promotes Cr enrichment in the scale and improves the oxidation resistance of steels. Usually, Si forms a Si oxide layer at the metal/oxide interface and the Si oxide layer prevents the oxidation of Fe. However, it is supposed that at low temperatures around 600°C, such as in this study, the Si oxide layer does not form and as a result, Si oxide mixes with the Cr rich oxide. The Cr content in the steel which produces excellent oxidation resistance can be less than 25%, if Si oxide is present in the oxide scale.

5 CONCLUSION

The oxidation behavior of Si bearing ferritic stainless steel was studied at around 600°C in a $H_2O + O_2$ atmosphere. It was concluded that:

1. The effect of O_2 in $O_2 + H_2O$ atmosphere: O_2 promotes the formation of M_2O_3-type oxide and prevents the occurrence of the nodular oxide scale. However, in the environment in which the nodular oxide scale forms, O_2 promotes the oxidation rate of steel.
2. The effect of H_2O in $O_2 + H_2O$ atmosphere: H_2O promotes the formation of M_3O_4-type oxide and the occurrence of the nodular oxide scale. It is supposed that the oxide scale formed in the H_2O bearing environment is apt to produce the defects.
3. In order to prevent the occurrence of the nodular oxide scale in a little O_2 and a greater quantity of H_2O environment, it is effective to increase the Cr content, and apply Si content in the steel. In addition, it is necessary that the Cr/Fe ratio in the Cr rich oxide layer is more than 3 and, as a result, the M_2O_3-type oxide is formed. Therefore, it is necessary that Cr and Si contents in the steel are Cr(%) + 4Si(%)27.

Acknowledgment

The authors are grateful to their laboratory colleagues, particularly to Dr. J.Murayama for the scale analysis, Mr. N.Maruyama for the oxidation tests, and Dr. S.Akiyama for the preparation of the test specimens.

References

1. D. Mortimer and W. B. A. Sharp: *Br. Corros. J.*, 1968, **3**, 61.
2. A. V.Sybolt: *Electrochem. Soc.*, 1960, **107**, 147.
3. G. P. Wazaldo and W.L.Pearl: *Corrosion*, 1965, 21, 355.
4. W. E. Ruther and G. Greenberg: *J. Electrochem. Soc.*, 1964, **111**, 1116.
5. C. T. Fujii and R. A. Meussner: *J. Electrochem. Soc.*, 1964, **111**, 1215.
6. T. Kawasaki, S. Sato and H. Ono: I.S.I.J., 1980, **65**, S344.
7. T. Kawasaki, S. Sato and H. Ono: *Corrosion Engineering*, Japan, 1982, **31**, 172.
8. H. Fujikawa, Y. Shida, N. Fujino and J. Murayama: *Corrosion Engineering*, Japan, 1982, **31**, 164.
9. J. Webber: *Corrosion Sci.*, 1976, **16**, 499.
10. I. Kvernes, M. Oliveira and P. Koffstad: *Corrosion Sci.*, 1977, **17**, 237.

1.1.4
The Influence of Processing Route on the Oxidation/Sulphidation of a Ti-48Al-2Nb-2Mn Intermetallic Alloy at 750 and 900 °C

H. L. Du[1], P. K. Datta[1], J. Leggett[2], J. R. Nicholls[2], J. C. Bryar[3] and M. H. Jacobs[3]

[1]SURFACE ENGINEERING RESEARCH GROUP, THE UNIVERSITY OF NORTHUMBRIA AT NEWCASTLE, NEWCASTLE UPON TYNE, NE1 8ST, UK
[2]SCHOOL OF INDUSTRIAL AND MANUFACTURING SCIENCE, THE UNIVERSITY OF CRANFIELD, BEDFORD, MK 43 0AL, UK
[3]IRC IN MATERIALS FOR HIGH PERFORMANCE APPLICATIONS, THE UNIVERSITY OF BIRMINGHAM, BIRMINGHAM, B15 2TT, UK

1 INTRODUCTION

Titanium aluminides-based intermetallics are potential candidate materials for high temperature applications due to their low density, good elevated temperature strength and excellent creep properties[1]. The intermetallic material, Ti-48Al-2Nb-2Mn, has been, and continues to be, the subject of wide ranging microstructural, processing, mechanical and environmental property studies within the Interdicipline Research Centre in Materials for High Performance Applications at the University of Birmingham[2-7]. This material has been developed to a stage where its environmental response to oxidising and sulphidising atmospheres at elevated temperatures needs to be characterised.

In this paper, the oxidation and sulphidation behaviour of the Ti–48Al–2Nb–2Mn fabricated by plasma melting, spray forming, spray forming plus hot isostatic pressing (HIPping) was studied in $H_2/H_2S/H_2O$ (high pS_2 and low pO_2) at 750 and 900°C. One of these process routes, plasma melted, has been further studied in an air/SO_2 environment (high pO_2 and low pS_2) also at 750 and 900°C.

Table 1 *Designation of the intermetallic materials by fabrication routes*

	Fabrication routes
Alloy 1	Spray formed – argon HIPped 1250°C
Alloy 2	Spray formed – vacuum HIPped 1250°C
Alloy 3	Spray formed – argon as-sprayed
Alloy 4	Spray formed – vacuum as-sprayed
Alloy 5	Plasma melted as-cast

2 EXPERIMENTAL

The designation of the materials processes by plasma melting, spray forming, spray forming plus hot isostatic pressing (HIPping) is given in Table 1. The materials were cut into 10 mm x 10 mm x 2 mm coupons using an ISOMET 2000 Precision Saw. The coupons were ground by 1200 grit SiC paper and were finally ultrasonically cleaned and degreased in acetone.

The oxidation/sulphidation experiments were carried out at 750 and 900°C for exposure periods of up to 168 hours. After the experimental rig was purged and flushed with N_2 and then H_2, a pre-mixed stream of H_2/H_2S was passed through a bubbler containing de-ionised water kept in a thermostat bath at 23°C. The pre-mixed H_2/H_2S finally entered the working tube carrying H_2O vapour. At 750°C, a pre-mixed 90%H_2/10%H_2S was used, thereby yielding an atmosphere of $pS_2\sim10^{-1}$ Pa and $pO_2\sim10^{-18}$ Pa, whilst at 900°C a gas mixture of 97.28%H_2/ 2.72%H_2S was pre-mixed and produced an atmosphere of $pS_2\sim10^{-1}$ Pa and $pO_2\sim10^{-14}$ Pa. The plasma melted sample was also exposed to air/SO_2 gas mixture containing approximately 500–700 ppm SO_2 at 750 and 900°C. The oxidation/sulphidation kinetics were determined by a discontinuous gravimetric method. The morphologies, compositions and phases present in the exposed specimens were characterised using scanning electron microscope (SEM), energy-dispersive analysis by X-ray (EDX), electron probe microanalysis (EPMA) and X-ray diffraction (XRD).

3 RESULTS AND DISCUSSION

3.1 Fabrication of Ti-48Al-2Nb-2Mn

The primary processing of the material is by plasma arc cold-hearth melting (PACHM) and the subsequent continuous casting of the melt to produce ingots.

3.1.1 Plasma Arc Cold-Hearth Melting (PACHM) of Ingots. Materials with very different microstructures were produced by secondary processing of the ingot material by cold-walled induction remelting, followed by bottom pouring of a molten stream which allows further processing by spray forming to produce ring-shaped preforms and curved sheets. These could further be processed by hot isostatically pressing (HIPping) to consolidate the material.

Figure 1 is a diagram of the Retech plasma arc cold-hearth melting furnace. The facility is capable of producing ingots of 100 mm and 150 mm diameter[2]. The cold hearth is 250 mm long, 150 mm wide and ~50 mm deep. During plasma melting, the furnace chamber was maintained at a positive pressure of ~ 1.1 bar by continuously purging with argon gas. Plasma melting was carried out with two 150 kW servo-hydraulic, computer-controlled transferred arc plasma torches which use helium as the plasma gas. One torch was positioned over the water-cooled copper hearth and serves two purposes; intermittently it was used to melt the feed stock which was fed in horizontally but, for the most part, it was used to maintain a liquid pool on the skull in the hearth. Additions of feed stock to the melt sequentially led to overflow of molten material via a notch into the water-cooled crucible with a retractable dove-tailed base, where the melt solidified to form an ingot. The second plasma torch was used to maintain a continuous liquid pool on top of the ingot (hot topping). Accurate control of the melt pool was essential if good quality ingots, defined in terms of uniformity of composition and microstructure were to be produced[3]. This liquid pool could be electromagnetically stirred by induction coils that surrounded the neck of the crucible.

For titanium aluminides the feed stock used was in compacted, elemental form composed of titanium sponge and elemental granules. The casting rate was ~ 0.5 kg per minute. Ingots up to 1.4 m long were cast.

3.1.2 Quality of Ti–48Al–2Nb–2Mn Titanium Aluminide *Ingots.* The quality of single melted plasma melted ingots was assessed in terms of macro-chemistry, macrostructure, macro-segregation and micro-segregation. For example, careful processing allows the aluminium concentration to be controlled within ±0.9 at% throughout in ingot[8].

The transverse and longitudinal optical, etched macrostructures for 100 mm diameter ingots are shown in Figures 2 (a) and 2 (b). There was a very narrow chill zone adjacent to the wall of the water-cooled crucible. Elsewhere, the macrostructure consisted of upward curving prior alpha columnar grains[4]. Figure 3 (a) shows the two phase ,$\alpha_2+\gamma$, internal structure of the columnar grains and Figure 3(b) shows that the prior alpha dendrites are cored, with aluminium-rich interdendritic pools.

3.1.3 Spray Forming-Centrifugal Spray Deposition (CSD). Plasma melted ingot material was clean remelted in a   radio frequency (RF) heated water-cooled copper induction crucible. Figure 4 shows a diagram of the segmented, bottom pour remelting facility. The induction crucible is capable of remelting 7000 cm^3 of material. Power for induction melting is supplied by a 350 kW, ~ 3 kHz supply; a second 86 kW, 10 kHz unit was independently used to heat the nozzle.

The bottom poured molten steam may be either gas atomised for spray forming (similar to Osprey processing and not shown in Figure 4) or centrifugal atomised - centrifugal spray deposition or CDS[5]. The fact that gas is not employed to effect the atomisation means that the CSD process can be operated either under an inert gas atmosphere or under low pressure, in a vacuum of ~ 200 mbar during spraying. Any porosity in deposits formed by LPCSD may be permanently removed by HIPping or mechanical deformation such as rolling[2]. Figure 4 shows

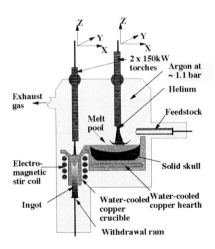

Figure 1 *Schematic diagram of plasma arc cold-hearth melting furnace*

the set up for low pressure centrifugal spray deposition, LPCSD. The spinning disc rotates at ~3000 rpm. The sprayed preform is deposited on the inside of a cylindrical steel substrate.

With this unique facility for centrifugal spray deposition it has been possible to prepare spray formed ingots of 400 mm diameter from Ti-48Al-2Nb-2Mn. Figure 5 illustrates typical microstructures obtained by spray forming Ti-48Al-2Nb-2Mn (a) in argon and (b) in vacuum[6]. For these trials a graphite bottom pour nozzle was used, with hole diameters in the range 2 to 5 mm. Detailed studies[7] of the microstructure of Ti-48Al-2Nb-2Mn material produced by LPCSD show that it consists of a lamellar grain structure with a fine dispersion of single phase gamma grains located primarily at the grain boundaries. Large columnar grains, characteristic of the ingot material, were completely absent.

3.2 Oxidation/Sulphidation of Ti-48Al-2Nb-2Mn in an $H_2/H_2S/H_2O$ Environment

3.2.1 Corrosion Kinetics. When the five alloys fabricated by different processing methods are exposed to $H_2/H_2S/H_2O$ environment at 750 and 900°C, their oxidation/sulphidation resistance is relatively insensitive to the processing routes, as shown in Figure 6. The oxidation/ sulphidation generally follows the parabolic rate law, However, the difference of weight gains

(a) Transverse section

(b) Longitudinal section

Figure 2 *The etched macrostructures for 100mm diameter ingots of Ti-48Al-2Nb-2Mn*

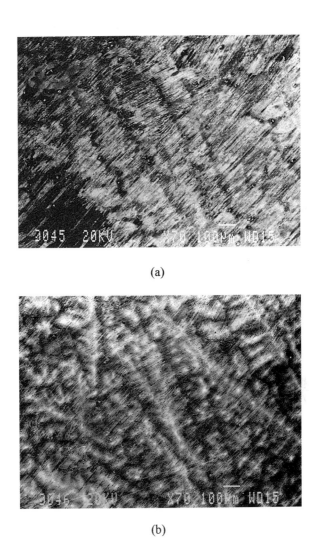

(a)

(b)

Figure 3 *(a) Optical micrograph showing lamelar of α-2 + γ phases, and (b) SEM primary backscatter image of dendrites, in a plasma melted Ti-48Al-2Nb-2Mn ingot*

between the maximum (Alloy 3) and minimum (Alloy 5) are very small and only 0.27 mg/cm^2. The oxidation/sulphidation kinetics of all alloys in H$_2$/H$_2$S/H$_2$O environment at 900°C also follow the parabolic rate law. The weight gains for all alloys are quite similar. That indicates that their oxidation/sulphidation resistance is not sensitive to the processing methods. However it must be pointed out that the weight gains for the Alloy 3 were persistently highest, which indicated the pores existing in this alloy during the fabrication did increase its corrosion rate by increasing exposure areas.

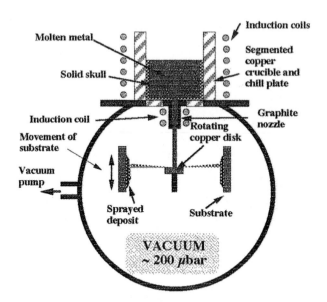

Figure 4 *Diagram of the facilities for clean remelting and low pressure centrifugal spray deposition (LPCSD)*

3.2.2 Corrosion Products. Figure 7 shows the micrographs for all five alloys after 72 hours exposure at 900°C. It is indicated that the scales formed on these alloys were compact and no spallation of the scale was observed. This is contrasted with similar alloy, Ti-46.6Al-2Mo-1.4Mn, on which the scales formed were very fragile and the spallation of the scale was observed even after one hour exposure[9]. There were no distinguished morphologies between these five alloys and, in another word, the morphologies were also insensitive to the processing routes, as reported in corrosion kinetic results. At 900°C and other exposure periods, all exposed alloys had similar scale morphologies. However scale spallation was observed after 168 hours exposure.

XRD analysis results showed that in the materials exposed for five hours, the intensive peak of γ-TiAl phase still existed. The formation of TiO_2 and Al_2O_3 on the surface of the samples was observed. XRD also showed the presence of TiS_2 and Al_2S_3. With increasing exposure time, the intensity of TiO_2 and Al_2O_3 increased.

SEM image and Digimaps for 168 hours exposed Alloy 5 in $H_2/H_2S/H_2O$ at 750°C are shown in Figure 8. Basically, the scale formed on the alloy consisted of three layers. The outer layer is TiO_2, beneath that is dominated by Al_2O_3. A complex sulphide layer consisting of TiS_2, Al_2S_3, Mn, Nb sulphides exists between the oxide layers and substrate. However the Mn and Nb sulphides were not detected by XRD, due probably to their small amount.

3.3 Oxidation/Sulphidation of Ti–48Al–2Nb–2Mn in an air/SO_2 Environment

3.3.1 Experimental Observations and Kinetics. The corrosion behaviour of Alloy 5 at 750 and 900°C is shown in Figure 9. Total mass gains of approximately 0.5 and 8.8 mg/cm², after 100 hours, were observed at 750 and 900°C respectively. Accelerated corrosion rates were

(a) In argon

(b) In vacuum

Figure 5 *Micrographs showing the through thickness microstructural variations in Ti-48Al-2Nb-2Mn rings produced by centrifugal spray deposition*

observed at both test temperatures. At 750°C parabolic kinetics were followed for 25 hours. An increase in corrosion rate was then observed between the 25th and 50th hours of exposure. Parabolic kinetics then continued at similar rates. At 900°C a much larger increase in the corrosion rate is observed between the 25th and 50th hour of exposure. This coincides with severe spalling effects, which indicated that non-protective scales formed at accelerated rates. Parabolic kinetics were followed up to 25 hours of exposure but due to the accelerated nature of corrosion after these breakaway kinetics, a combination of parabolic and linear kinetics was followed.

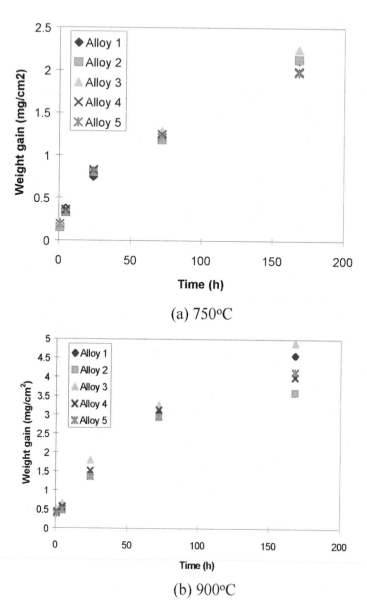

(a) 750°C

(b) 900°C

Figure 6 *Oxidation/sulphidation kinetics of Alloys 1-5 exposed to $H_2/H_2S/H_2O$*

The surface morphologies seen after 100 hours at 750 and 900°C are shown in Figure 10. At 750°C a dispersion of manganese and titanium sulphide particles was identified, using EPMA, on a predominantly porous rutile surface layer. The surface microstructure at 900°C differed in that the rutile grains were much larger and there were no sulphide particles in evidence. Because grains of manganese oxide were also identified it would appear that the sulphide particles seen at 750°C are not stable and react with oxygen to form oxides after longer

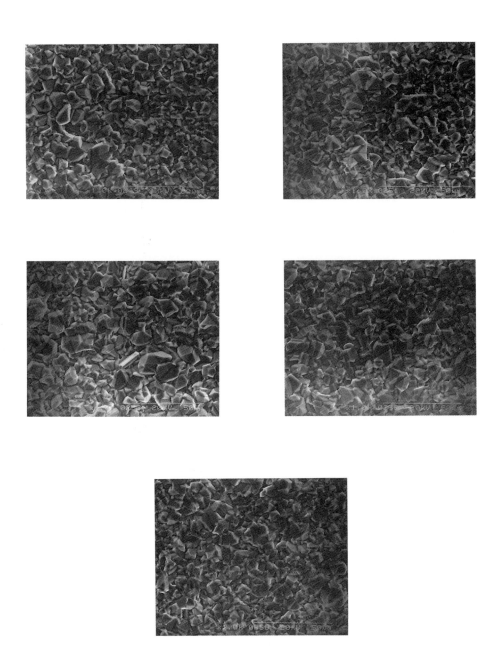

Figure 7 *SEM micrographs for exposed Alloys 1–5 in $H_2/H_2S/H_2O$ at 900°C for 72 hours*

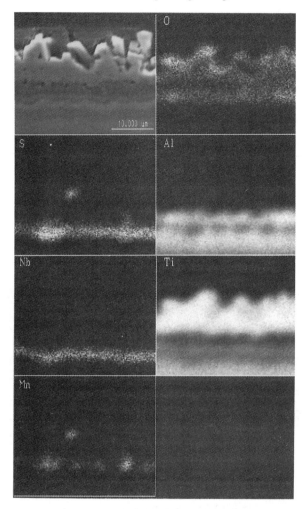

Figure 8 *SEM image and digimaps showing compositional profiles of Alloy 5 after 168 hours exposure in $H_2/H_2S/H_2O$ at 750°C*

periods of time and at higher temperatures. Figure 11 shows an example of the transverse section of the scales which develop. In this case, 100 hours at 900°C. The original scales have spalled and new scales reformed. Spalling is observed after a critical scale thickness is reached and thus the scale fracture stress exceeded.

Element analysis was used to identify the scales developed. Alumina particles were identified below the rutile/gas interface but still within the outer rutile layer. Below the outer rutile plus manganese oxide surface layer a discontinuous alumina-rich layer formed with areas rich in titanium. Beneath this layer a rutile layer was identified containing alumina and niobium which may have been in the form of particles. Near the scale/substrate interface a sulphur/niobium rich layer existed, as well as a high proportion of titanium.

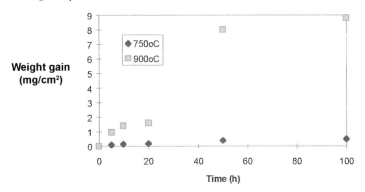

Figure 9 *Oxidation/sulphidation kinetics of Alloy 5 exposed to air/SO$_2$ at 750 and 900°C*

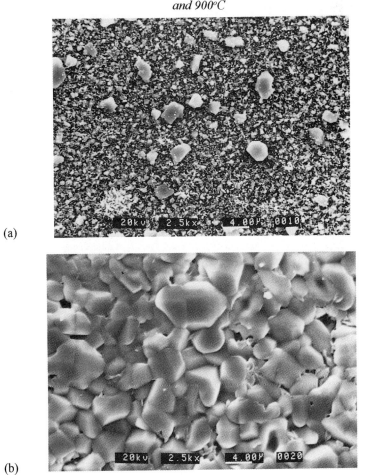

(a)

(b)

Figure 10 *SEM micrographs of Alloy 5 after 100 hours exposure in air/SO$_2$*

Figure 11 *Cross-sectioned morphology of exposed Alloy 5 in air/SO₂ at 900°C for 100 hours*

3.3.2 Oxidation/Sulphidation Mechanism. A three stage oxidation/sulphidation mechanism has been proposed to explain scale formation on the Ti-48Al-2Nb-2Mn alloy exposed to a high pO_2 and low pS_2 environment, and is shown in Figure 12.

Initially a porous rutile plus alumina surface layer forms, with a dispersion of manganese and titanium sulphides forming on this layer. In the second stage the rutile layer thickens and incorporates the sulphides. At the same time the sulphides become unstable and react with oxygen in the atmosphere. This releases sulphur, which along with sulphur from the atmosphere, diffuses through the rutile scale, where it reacts with outwardly diffusing cations to form a continuous sulphide layer. The third stage involves rapid outward diffusion of titanium and aluminium through the sulphide scale. This results in a discontinuous alumina and rutile layer just beyond the sulphide layer. However, aluminium continues to diffuse outward where there is a competition with titanium to form oxides. Although two different layers develop beneath the surface rutile layer, both contain areas rich in the other element. Hence a non-protective scale develops and a continuous alumina layer is prevented from forming.

4 CONCLUSIONS

1. The plasma melting, spray forming and spray forming plus HIPping were successfully employed in producing Ti-48Al-2Nb-2Mn intermetallics.

Stage 1:

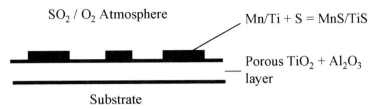

SO$_2$ / O$_2$ Atmosphere

Mn/Ti + S = MnS/TiS

Porous TiO$_2$ + Al$_2$O$_3$ layer

Substrate

Stage 2

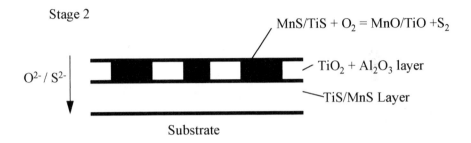

MnS/TiS + O$_2$ = MnO/TiO +S$_2$

TiO$_2$ + Al$_2$O$_3$ layer

TiS/MnS Layer

O^{2-} / S^{2-}

Substrate

Stage 3

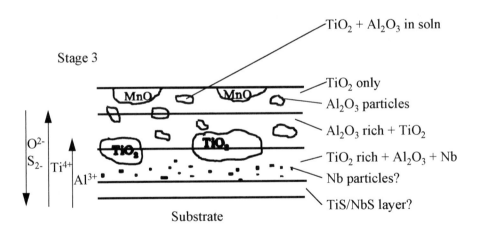

TiO$_2$ + Al$_2$O$_3$ in soln

TiO$_2$ only

Al$_2$O$_3$ particles

Al$_2$O$_3$ rich + TiO$_2$

TiO$_2$ rich + Al$_2$O$_3$ + Nb

Nb particles?

TiS/NbS layer?

MnO   MnO

TiO$_2$ TiO$_2$

O^{2-}
S$_{2-}$ Ti^{4+}
       Al^{3+}

Substrate

Figure 12 *Oxidation/sulphidation mechanism of Ti-48Al-2Nb-2Mn*

2. The oxidation/sulphidation in H$_2$/H$_2$S/H$_2$O (high pS$_2$ and low pO$_2$) and air/SO$_2$ (high pO$_2$ and low pS$_2$) at 750 and 900°C generally followed parabolic rate law. However, breakaway kinetics were observed after long-term exposure at 900°C.
3. The corrosion resistance of Ti-48Al-2Nb-2Mn was relatively insensitive to the processing routes.

References

1. D. M. Dimiduk, D. B. Miracle and C. H. Ward, *Mater. Sci. Technol.*, 1992, **8**, 367.
2. M. H. Jacobs and J. M. Young, Primary plasma melting of aerospace alloys and their Secondary Processing by Spray Forming and Gas Atomisation, in Proc. of the 1994 Int. Symp. on Liquid Metal Processing and Casting (Ed. A Mitchell and J. Fernihough), Santa Fe, New Mexico, 1994, 43–53, American Vacuum Society.
3. R. M. Ward, A. E. Fellows, T. P. Johnson, J. M. Young and M. H. Jacobs, Plasma Melting: Intelligent Control Strategies, *J de Physique IV*, 1993, Colloque C7, supplement au J de Physique III, **3**, 823–828.
4. T. P. Johnson, J. M. Young, R. M. Ward and M. H. Jacobs, Influence of Melting Conditions on Structure and Quality of Plasma Melted Titanium Aluminides, Proc. Symp. On Structural Intermetallics, Seven Spring, TMS, 159-165, 1994.
5. T. P. Johnson, M.H. Jacobs R.M. Ward and J.M. Young, Spray forming of alloys for high performance application, Proc. 2nd Int. Conf. On Spray Forming, Swansea, UK, 129–139, Ed. J.V. Wood, Woodhead Publishing Ltd., 1993.
6. A. L. Dowson, M. A. Duggan, Y. Y. Zhao, M. H. Jacobs and J.M. Young, Microstructure and thermo-mechanical processing response of spray formed Ti alloys produced by centrifugal spray deposition, 1996 World Congress on Powder Metallurgy & Particulate Materials, Washington DC, USA, Metal Powders Industries Federation.
7. M. H. Jacobs, M. A. Ashworth, T. P. Johnson and J. M. Young, Production, Processing and Characterisation of Clean-melted, Gas-atomised titanium powders, Proc. of 3rd Int. Conf. on Powder Metallurgy, Aerospace, Defence and Demanding Applications, MPIF, Ed. F.H. Foes, 233–240, 1993.
8. A. L. Dowson, T. P. Johnson and R. M. Ward, IRC in Materials, private communication.
9. H. L. Du, P. K. Datta and S. K. Hwang, High temperature corrosion behaviour of Ti–46.6Al–2Mo–1.4Mn in environments of low oxygen and high sulphur potentials at 750 and 900°C, presented at 4th Int. Symp. on High Temperature Corrosion and Protection of Materials, France, May 1996.

1.1.5
Sulphidation Properties of Ti-Al Intermetallic Compounds at High Temperatures in H$_2$S-H$_2$ Atmospheres

Takayuki Yoshioka[1] and Toshio Narita[2]

[1]GRADUATE SCHOOL OF HOKKAIDO UNIVERSITY

[2]RESEARCH GROUP OF INTERFACE CONTROL ENGINEERING, GRADUATE SCHOOL OF ENGINEERING, HOKKAIDO UNIVERSITY, 060 SAPPORO, JAPAN

1 INTRODUCTION

Intermetallic Ti-Al compounds are attracting interest for high temperature applications[1-3] because of their high specific strength at elevated temperatures. To use these alloys in practical applications, their mechanical and anti-oxidation properties have to be better understood.

A number of investigations has been carried out to improve the mechanical properties of Ti-Al alloys at room temperature[4-6] and ductility has also been improved. The anti-oxidation properties of Ti-Al alloys are still unsatisfactory due to rapid oxidation with formation of oxide scales of TiO_2 and Al_2O_3. There has been a number of investigations dealing with the oxidation mechanism[7-11], and a number of ways to improve oxidation resistance[12-16] has been reported.

There is little information on the sulphidation properties of Ti-Al alloys, while sulphidation will become important when the alloys are subjected to aggressive environments as both Ti and Al are very reactive with sulphur.

In this investigation, the sulphidation behaviour of Ti_3Al, TiAl, and $TiAl_3$ alloys as well as Ti metal, for comparison purposes, was investigated at temperatures of 1073 and 1173K under various sulphur partial pressures in H$_2$S-H$_2$ gas mixtures.

2 EXPERIMENTAL

2.1 Specimens

Intermetallic Ti-Al compounds (Ti_3Al, TiAl and $TiAl_3$) were made by argon-arc melting using Ti and Al as starting materials, and then annealed in vacuum at 1373K for 86.4ks. Titanium plate with a commercial grade purity of 99.5 at% Ti was used for comparison purposes. The chemical compositions of Ti-Al alloys are summarized in Table 1. From microstructural observations and compositional analyses, it was found that both Ti_3Al and TiAl are in the single phase regions in the Ti-Al alloy system[17], while the $TiAl_3$ alloy contains small amounts of $TiAl_2$ phase and casting faults such as voids.

The specimen size was $10 \times 10 \times 1$ mm^3, and the specimen surface was abraded with SiC papers down to #1500 and then polished using 3-micron diamond paste to a mirror-like finish.

Table 1 *Chemical composition of the Ti-Al alloys, (at%)*

	Al	Cu	W	N	O	C	Ti
Ti₃Al	25.4	0.007	0.002	0.02	0.2	0.04	remaining
TiAl	50.4	0.006	0.002	0.01	0.1	0.06	remaining
TiAl₃	74.4	0.005	0.002	0.007	0.04	0.1	remaining
Ti	99.95%		grade		commercial		sheet

Prior to the sulphidation experiments the specimens were ultrasonically cleaned in methanol-benzene solution.

2.2 Sulphidation Conditions and Apparatus

An H_2S and H_2 gas mixture was used as the corrosion atmosphere. The gas mixture was prepared from pure H_2S (4N grade) and H_2 (6N grade) gases using commercially available gas (Daido–Hokusan 1Ad., Sapporo). Minute adjustment of the gas composition was made by adding H_2 gas into the pre-mixed H_2S-H_2 gases. In this investigation the gas mixture was introduced into a reaction tube without any further purification treatment. Gas composition at room temperature and their sulphur partial pressures at the reaction temperatures are given in Table 2.

Figure 1 shows a schematic representation of the sulphidation apparatus. The specimen was suspended from a quartz spring with a platinum wire and then settled in a position near the top of a thermocouple, as shown in Figure 1. The H_2S/H_2 gas mixture was introduced into the reaction tube after several argon-gas flushings, and then sulphidation was commenced by raising the furnace to the desired position, sulphidation was stopped by lowering the furnace. The gas flow rate in the reaction tube is approximately 2cm/s at the sulphidation temperatures. A batch-type measurement was adopted for measurements of corrosion amounts for each sulphidation duration. The surface morphology and cross-sectional micro-structures of the sulphide scale and alloy substrates were observed using a SEM-EDX (scanning electron microscope equipped with an energy dispersive analyser). The sulphides were identified by XRDA (X-Ray Diffraction analysis) and concentration profiles were measured using EPMA (electron probe micro-analysis).

Table2 *Sulphur partial pressures in the H_2S-H_2 gas mixtures at 1073K and 1173K*

At 1073K	0.1 vol% H₂S	2.2 vol% H₂S	9.7 vol% H₂S
log pS2	−4.7	−2.0	−0.7

At 1173K	0.1 vol% H₂S	2.1 vol% H₂S	9.7 vol% H₂S
log pS2	−3.9	−1.3	0.1

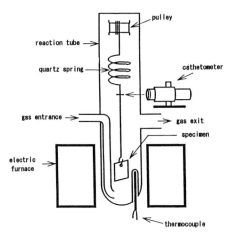

Figure 1 *Schematic diagram of sulphidation apparatus*

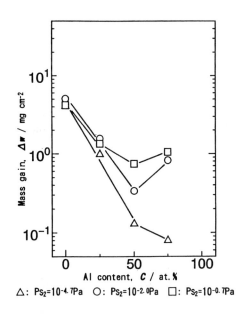

△: Ps₂=10⁻⁴·⁷Pa   O: Ps₂=10⁻²·⁰Pa   □: Ps₂=10⁻⁰·⁷Pa

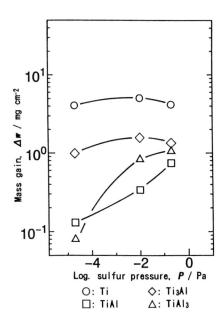

O: Ti   ◇: Ti₃Al
□: TiAl   △: TiAl₃

Figure 2 *Sulphidation amounts vs. aluminium content of alloys sulphidized at 1073K for 86.4ks at various sulphur pressures*

Figure 3 *Sulphur pressure vs. sulphidation amounts of alloys sulphidized at 1073K for 86.4ks*

3 RESULTS AND DISCUSSION

3.1 Sulphidation Amounts for 86.4ks

Figures 2 and 3 show sulphidation amounts plotted against alloy compositions and sulphur pressures for Ti, Ti$_3$Al, TiAl, and TiAl$_3$, sulphidized at 1073K for 86.4ks under sulphur pressures of $10^{-4.7}$, $10^{-2.0}$ and $10^{-0.7}$Pa. At each sulphur pressure, the sulphidation amounts decreased steeply with increasing Al contents from pure Ti to a TiAl alloy and they level out for the TiAl$_3$ alloy. There is some scattering of the TiAl$_3$ data due to voids in the TiAl$_3$ alloy surface.

The sulphur pressure dependence of the sulphidation amounts show that pure Ti and the Ti$_3$Al alloy are almost independent of sulphur pressure, while sulphidation amounts for the TiAl and TiAl$_3$ alloys decreased rapidly with decreasing sulphur pressures.

Figure 4 shows sulphidation amounts of Ti, Ti$_3$Al, TiAl and TiAl$_3$ after sulphidation at 1173K for 86.4ks under sulphur pressures of $10^{-3.9}$, $10^{-1.3}$, and $10^{0.1}$Pa. Sulphidation amounts decreased steeply with increasing Al content at $10^{-3.9}$Pa sulphur pressure, while sulphidation at $10^{-1.3}$ and $10^{0.1}$Pa decreased slowly from pure Ti to the TiAl alloy and then decreased rapidly with the TiAl$_3$ alloy. Accordingly, the sulphur pressure dependence increased from pure Ti to Ti$_3$Al and TiAl alloys in this sequence, as shown in Figure 5.

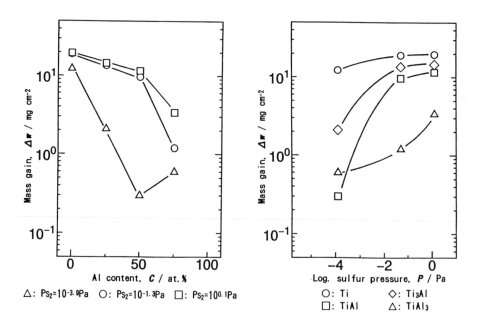

Figure 4 *Sulphidation amounts vs. aluminum content of alloys surlphidized at 1173K for 86.4ks at various sulphur pressures.*

Figure 5 *Sulphur pressure vs. sulphidization amounts of alloys sulphidized at 1173K for 86.4ks.*

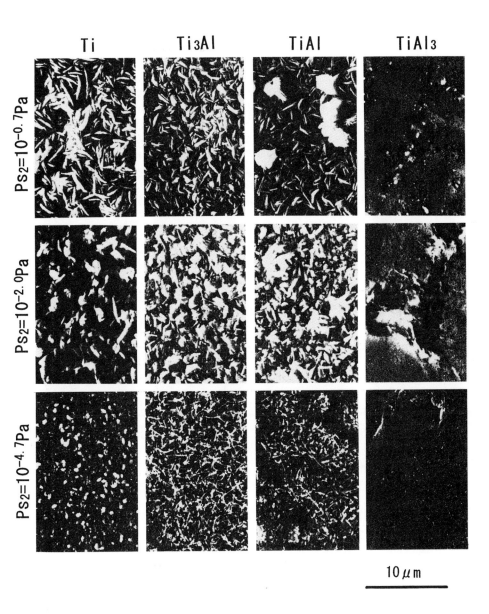

Figure 6 *Surface morphologies of as-grown sulphide scales formed on alloys sulphidized at 1073k for 86.4ks at various sulphur pressures*

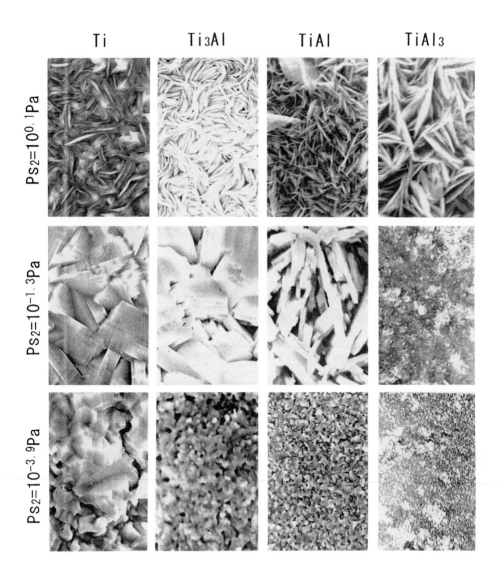

Figure 7 *Surface morphologies of as-grown sulphide scales formed on alloys sulphidized at 1173K for 86.4ks at various sulphur pressures*

3.2 Surface morphologies of the sulphide scales

Figure 6 shows the surface morphologies of the as-grown sulphide scales formed on Ti, Ti_3Al, TiAl, and $TiAl_3$ sulphidized for 86.4ks under various sulphur pressures at 1073K. For the pure Ti, Ti_3Al and TiAl alloys, the surface morphologies showed a fibrous structure, which became an agglomeration of small grains with decreasing sulphur pressures. For the $TiAl_3$ alloy, the corrosion products with blue and light brown interference colours were thin and scratches are observed. Figure 6 demonstrates that sulphides grow relatively thick near the voids, which were formed during the casting of the alloy ingot.

Figure 7 shows surface morphologies of as-grown sulphide scales formed on Ti, Ti_3Al, TiAl and $TiAl_3$ sulphidized for 86.4ks at various sulphur pressures at 1173K. These sulphide morphologies could be divided into three groups – a fibrous structure at $10^{0.1}$ Pa sulphur pressure, a faceted structure for Ti, Ti_3Al and TiAl alloys at $10^{-1.3}$ Pa and Ti at a $10^{-3.9}$ Pa sulphur pressures; and a very fine structure for the $TiAl_3$ alloy at $10^{-1.3}$ and $10^{-3.9}$ Pa sulphur pressures as well as for the Ti_3Al and TiAl alloys at $10^{-3.9}$ Pa sulphur pressure.

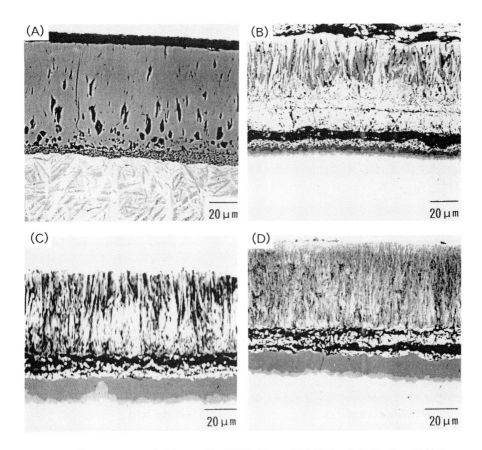

Figure 8 *Cross-section of: (a) pure Ti, (b) Ti_3Al, and (c) TiAl sulphidized at $10^{-1.3}$ Pa sulphur pressure at 1173K for 86.4ks and (d) TiAl, Sulphidized at $10^{0.1}$ Pa sulphur pressure at 1173K for 86.4ks*

3.3 Cross-sectional Microstructures

As shown in Figures 6 and 7, the surface morphologies could be classified into three groups depending on the alloy composition and sulphur pressure. Figure 8 shows cross-sectional microstructures of Ti, Ti_3Al, and TiAl after sulphidation at 1173K for 86.4ks at $10^{-1.3}$ Pa sulphur pressure and TiAl after sulphidation at 1173K for 86.4ks at $10^{0.1}$ Pa sulphur pressure. When sulphidized at a $10^{-1.3}$ Pa sulphur pressure, the sulphide scale formed on pure Ti was mainly an outer layer and a thin porous inner layer, whereas the sulphide scale with fibrous structure was found in the inner/intermediate/outer triplex structure on the Ti_3Al alloy and the inner/outer duplex structure on the TiAl alloy. Further, it was observed that a new phase was formed on the alloy surface. With increasing sulphur pressure, the scale structure formed on the TiAl alloy became very fine.

3.4 Corrosion Products Identified by XRD

X-ray diffraction patterns of the successively polished sections of the scale and alloy substrate of the sulphidized TiAl alloy were measured. A mixture of TiS_2, Ti_3S_4, TiS, and Al_2S_3 was identified in the outer scale, while TiS with small amounts of Ti_2S was observed in the inner scale. Although there are unidentified diffraction peaks in the XRD spectra, it could be concluded that the TiS_2 peaks correspond to the sulphide with fibrous structure, the Ti_3S_4 to the faceted structure, and the TiS to the finely grained structure.

As demonstrated in Figures 6 and 7, the TiAl alloy formed a thin scale at relatively low sulphur pressures, and the XRD analysis showed a formation of Al_2O_3, in addition to indications of sulphides and sulphur detected by the EPMA analysis. The formation of Al_2O_3 could be explained by assuming water vapour as impurity in the H_2S/H_2 gas mixture. For an atmosphere containing 1 ppm H_2O, the oxygen partial pressure is estimated to be approximately 10^{-23} Pa, high enough to form Al_2O_3, as the dissociation pressure of Al_2O_3 is close to 10^{-31} Pa at 1173K.

4 CONCLUSIONS

Intermetallic Ti-Al compounds Ti_3Al, TiAl, and $TiAl_3$ as well as pure Ti metal were sulphidized for 86.4ks at 1073 and 1173K under various sulphur pressures in H_2S/H_2 gas mixtures. The results obtained could be summarized as follows:

1. Sulphidation amounts at 1073K decreased steeply with increasing Al content in the sequence Ti, Ti_3Al and TiAl, while it was similar for the TiAl and $TiAl_3$ alloys. At 1173K the sulphidation amounts decreased slowly from Ti to TiAl and then rapidly toward the $TiAl_3$ alloy.
2. The XRD analysis identified a mixture of titanium and aluminium sulphide in the outer layer, whereas there was only titanium sulphide in the inner layer.
3. The morphologies and XRD analysis showed the TiS_2 to correspond to the fibrous structure sulphide, the Ti_3S_4 corresponded to the faceted structure, and TiS to the finely grained structure.

4. The TiAl$_3$ and TiAl$_2$ layers were formed on the subsurface of Ti$_3$Al and TiAl alloys by preferential sulphidation of titanium.

Acknowledgments

The authors wish to thank Dr. Y. Shida of Sumitomo Metal Industry Ltd. for supplying the Ti-Al intermetallic compounds.

References

1. S. M. L.Sastry and H. A. Lipsitt, Titanium '80, Ed. by H.Kimura and O. Izumi, The Metallurgical Society of AIME, 1980, 1231.
2. M. Hunt, *Material Engineering*, 1990, 35.
3. M.Yamaguchi, S. R. Nishitani and Y .Shirai, High Temperature Aluminides & Intermetallics, Ed. by S.H.Whang, C.T.Liu, D.P.Pope, and J.O.Stiegler, The Minerals, Metals & Materials Society, 1990, p. 63.
4. S. A. Court, V. K. Vasudevan and H. L. Fraser, *Philosophical Magazine A*, 1990, **61,** 141.
5. H. Inui, A. Nakamura, M. H. Oh and M. Yamaguchi, 'Intermetallic Compounds', Ed. by O. Izumi, The Japan Institute of Metals, 1991, p. 495.
6. T. Kawabata, T. Tamura and O. Izumi, *Metallurgical Transactions A*, 1993, **24A,** 141.
7. N. S. Choudhury, H. C. Graham, and J. W. Hinze, ' Properties of High Temperature Alloys with Emphasis on Environmental Effects', Ed. by Z.A.Foroulis and F.S.Pettit, The Electrochemical Society, 1976, p. 668.
8. K. L. Luthra, *Oxidation of Metals*, 1991, **36,** 475.
9. H. Anada and Y. Shida, 'Intermetallic Compounds', Ed. by O.Izumi, The Japan Institute of Metals, 1991, p. 731.
10. K. Kasahara and M. Takeyama, *J. Japan Inst. Metals*, 1993, **57,** 1288.
11. A.Rahmel, W. J. Quadakkers and M. Schijtze, *Materials and Corrosion*, 1995, **469,** 271.
12. Y. Shida and H. Anada, *J. Japan Inst. Metals*, 1994, **58,** 754.
13. S. Taniguchi, T. Shibata and S.Sakon, 'Intermetallic Compounds', Ed. by O. Izumi, The Japan Institute of Metals, 1991, p. 719.
14. M. Yoshihara, K .Miura and R. Tanaka, Report of the 123rd **JSPS., 327,** 1991, p. 337.
15. A. Takei and A.Ishida, 'High Temperature Corrosion of Advanced Materials and Protective Coatings', Ed. by Y.Saito, B. Önay, and T. Maruyama, Tokyo Institute of Technology, 1992, p. 317.
16. J. L. Smialek, Corrosion Science, 1993, **35,** 1199.
17. Binary Alloy Phase Diagrams, ASM International, vol. l, 225.

1.1.6

High Temperature Corrosion of Electroless Nickel Film Formed on Iron in Flowing Chlorine

T. Jiangping, L. Jian and L. Maoshen

DEPARTMENT OF MATERIALS SCIENCE AND ENGINEERING, ZHEJIANG UNIVERSITY, HANGZHOU, 310027, CHINA

1 INTRODUCTION

The corrosion of many metals in chlorination environments is greatly accelerated due to the formation of chlorides, which generally have low melting points and high vapour pressures[1-4]. The upper temperatures of these metals for continuous service in chlorine containing environments are limited[5-6]. Resistance to halogen corrosion at high temperature requires the formation and maintenance of a stable coherent and adherent surface layer. The rates of chlorination of metals are decreased by protective layers which form on the surfaces of the metals[7].

While iron is the least expensive structural metal, it cannot be used without alloying in applications that require exposure to chlorine containing environments at elevated temperaures because iron chlorides volatolise rapidly above 300ºC[8,9]. Several attempts have been made to improve the resistance against chlorination[10,11].

Among metals, nickel has some importance because of its comparable melting point and low vapour pressure of nickel chloride[12]. Electroless nickel deposits have some important advantages over other types of nickel coatings[13]. Using the process of electroless deposition it is possible to obtain thick, uniform relatively pure nickel deposits without the problems associated with electroplating. Electroless nickel film deposits on the surface of iron can act as a diffusion barrier, preventing the inward diffusion of chlorine through the film and outward vapour transport.

In this work, the chlorination behaviour of thick nickel film on iron by electroless method has been investigated. The damage mechanisms of the electroless nickel film by chlorine have also been discussed.

2 EXPERIMENTAL PROCEDURES

The pure iron (99.99%) substrates ($10 \times 10 \times 1.7$ mm) were first washed in a 0. 1 % neutral detergent solution. The electroless nickel bath composition was:

Nickel acetate	60g/L
Glycolic acid	60g/L

Figure 1 *Cross-section of micrograph of electroless nickel film on iron*

Tetrasodium EDTA 25g/L
Hydrazine   100ml/L

The EDTA was chosen because it produced a stabilizing effect in electroless solution. For this series of experiments, the temperature was maintained at about 90°C, and pH at about 11.0. The thickness of the deposits was checked by weighing the panels before and after plating. By using the formulation at the operating conditions, deposits up to 12 μm thick were obtained.

Adhesion of the nickel deposits was good on iron. When plated panels were bent 180 degrees, cracks appeared in the deposits but there was no delamination from the base metal. Finally the specimens were subjected to heat treatment in a vacuum at 250°C for one hour.

The thermogravimetric experiments were carried out in a gas supply system. Water in chlorine gas was absorbed by H_2SO_4. The superficial velocity of chlorine flowing through the reaction tube was 1.5m/s. The temperature in the reaction zone was maintained at 400°C ± 3°C. The weight change of the specimens during corrosion was measured by Cahn 1000 balance. Before each experiment, the apparatus was purged with nitrogen to remove residual chlorine, evacuated to about 10^{-3} Pa and then heated to the experimental temperature. Following this, the chlorine gas was introduced.

After each experiment, the exposed specimens were examined by SEM and EDS. The corrosion products remaining on the surfaces of the specimens were analyzed using XRD.

3 RESULTS AND DISCUSSION

3.1 Electroless Nickel Deposit

Chemical analysis of the deposits shows 99.0 percent nickel. Spectrographic analysis showed traces of aluminum, calcium, cobalt, copper, iron, lead, magnesium, silicon and silver.

As shown in Figure 1, structure of electroless nickel deposits is extremely fine grained. This result was verified by X-ray diffraction, which showed line broadening typical of small particle size. Density measurement after heat treatment was 8.49g/cm^3, indicating the presence of voids in the nickel deposits.

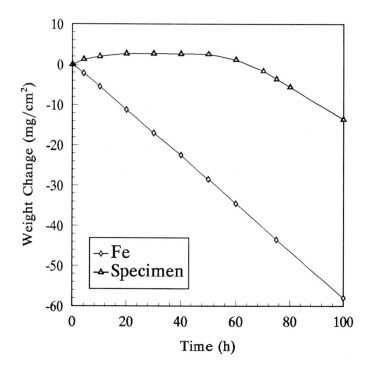

Figure 2 *Weight change results for the specimen exposed to Cl_2 at 400°C*

3.2 Chlorination Kinetics

Thermogravimetric tests have been performed at 400°C in flowing chlorine gas. The specimen surfaces were analyzed by X-ray diffraction after the corrosion test. The diffraction patterns corresponded to $NiCl_2$, FCC Ni and BCC Fe. Only a small amount of $FeCl_3$ and $FeCl_2$ at the beginning of exposure and a large amount of $FeCl_3$ and $FeCl_2$ after exposure to chlorine gas for a long time were detected. Figure 2 shows the results of the weight change measurements of the specimens. This figure also shows the thermogravimetric result of iron in chlorine gas at 400°C for reference.

In each case the mass of iron decreases with time according to an approximately linear rate law. However, the kinetics processes of nickel deposition specimens show different features. The weight change of the specimens versus time represents three stage:

1. an increase in weight according to a parabolic rate law,
2. no change or a little decrease in weight,
3. a linear decrease in weight.

Thermochemical considerations regarding the reaction of electroless nickel deposits on iron with chlorine gas at 400°C indicate that the following metal chlorides are of significance: $FeCl_2$, $FeCl_3$ and $NiCl_2$. For the electroless nickel deposit, plots of weight change versus time can be adequately expressed as:

$$\frac{d}{dt}(wt)_{Ni} = -\frac{d}{dt}(Ni) + \frac{d}{dt}(NiCl_2)_F - \frac{d}{dt}(NiCl_2)_V \qquad \{1\}$$

where $\frac{d}{dt}(Ni)$, $\frac{d}{dt}(NiCl_2)_F$ and $\frac{d}{dt}(NiCl_2)_V$ are the rate of weight loss of metallic nickel, the rate of formation of $NiCl_2$ and the rate of volatilization of $NiCl_2$, respectively.

At 400°C, where an appreciable volatilization of $NiCl_2$ was not observed, the volatilization of $NiCl_2$ can be ignored. Then equation $\{1\}$ can be written as:

$$\frac{d}{dt}(wt)_{Ni} = -\frac{d}{dt}(Ni) + \frac{d}{dt}(NiCl_2)_F \qquad \{2\}$$

For the iron substrate, the weight change versus time can also be expressed as:

$$\frac{d}{dt}(wt)_{FE} = -\frac{d}{dt}(Fe) + \frac{d}{dt}(FeCl_3 + FeCl_2)_F - \frac{d}{dt}(FeCl_3 + FeCl_2)_V \qquad \{3\}$$

where $\frac{d}{dt}(Fe)$, $\frac{d}{dt}(FeCl_3 + FeCl_2)_F$ and $\frac{d}{dt}(FeCl_3 + FeCl_2)_V$ are the rate of weight loss of metallic iron, the rate of formation of iron chlorides and the rate of volatilization of iron chlorides, respectively.

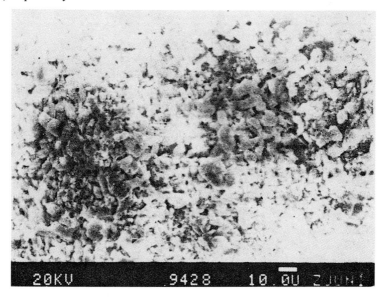

Figure 3 *Scanning electron micrograph of the corroded surface formed on the specimen after 2 hours of exposure at 400°C*

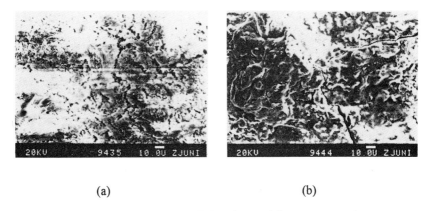

(a)   (b)

Figure 4 *SEM micrographs of the corroded surfaces of the specimens after exposure at 400ºC*

The reaction between Fe and Cl_2 consists of the following steps:

1. transfer of chlorine toward the specimen surface across the laminar boundary layer,
2. diffusion of elemental chlorine across the film,
3. formation of solid chlorides,
4. formation of volatile chlorides and
5. transfer of volatile chlorides away form the interface through the film.

The chlorination resistance of iron has been considerably improved due to electroless nickel film.

During the first stage of chlorination, the thick, uniform nickel film on iron blocked chlorine directly reacts with iron. The corrosion resulted from the diffusion of elemental chlorine and the formation of $NiCl_2$, $FeCl_2$ and $FeCl_3$. The rate of formation of iron chlorides was controlled by the diffusion of elemental chlorine through the film. A uniform surface morphology of specimen shown in Figure 3. Although XRD results demonstrated that a majority of the corrosion products formed during the exposures, a substantial quantity of condensed corrosion product, $NiCl_2$ was found on the surfaces of the specimens after the experiments. No iron chlorides were detected by EDS microanalysts. It indicated that the uniform nickel film on iron and condensed nickel chloride formed on the nickel film hindered the transport of volatile iron chlorides form the film-metal interface to gas phase. The formation of chlorides was predominant during this stage. Because voids in nickel deposit can be a path for short-circuit diffusion, the rate of formation of dense iron chlorides was rapid. Then the specimens showed a slightly more net increase in weight over the same time than given by a parabolic rate law.

During the second stage of chlorination the nickel film suffered damage to some extent. Figure 4 shows the SEM micrographs of surfaces of the specimens after exposure. The film shows pores and cracks, permitting the metal to be attacked by chlorine. Dense iron chloride deposits were detected between the nickel film and the metal after the corrosion of these specimens, indicating that chlorine was able to penetrate through this nickel film. Above 316ºC, $FeCl_3$ exists in the molten state[14]. In the temperature investigated, some volatile chlorides can transport from the interface through the pores and cracks to the gas phase. In order to

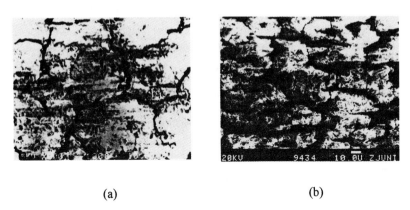

(a) (b)

Figure 5 *SEM micrographs of uncorroded surfaces of the specimens in chlorine gas at 400°C showing (a) crack and (b) blister*

follow the change in corrosion process with time, the weight change of the specimen was measured by a quartz microbalance at a constant time interval during exposure to the chlorine gas. The specimen weight increased by the formation of chlorides but decreased by volatilization of chlorides. Under steady state conditions, the rates of formation and volatilization of chlorides are equal. Then the weight of the specimen tends to remain constant or show a slight decrease after further exposure to chlorine gas.

After long-term exposure to chlorine gas, a nearly linear decrease in the specimen is observed. During this stage, the protective nickel film on iron was completely destroyed. A large amount of volatile or molten ferric chloride between the film and metal may be responsible for blistering of the film and the loss of adhesion. The porous film could not hinder the transport of volatile ferric chloride form the film/metal interface to the gas phase. The corrosion processes were controlled by the volatilization of ferric chloride and the weights of the specimens decrease with time during the third stage according to an approximately linear rate law. Because of the presence of nickel film and condensed $NiCl_2$ formed on the specimen surface, the rate constant of the specimen was smaller compared to that of iron chlorination.

3.3 Damage of Electroless Nickel Film by Chlorine

When chlorine is able to penetrate the film/ metal interface, the metal chlorides will be produced in this region. At the film-metal interface, the following reactions occur:

$$Fe(s) + Cl_2(g) = FeCl_2(s)$$
$$FeCl_2(s) + 1/2Cl_2(g) = FeCl_3(s)$$
$$FeCl_2(s) = FeCl_2(v)$$
$$FeCl_3(s) = FeCl_3(v)$$

In comparison with the melting points and the equilibrium vapour pressures of iron chlorides at 400°C, ferric chloride becomes molten and has very high vapour pressure. With increasing exposure time, substantial quantities of iron chlorides are produced near the corroding surface.

Consequently, the formation of very- volatile ferric chloride can generate sufficient pressure to produce mechanical lifting of the film. Molten chlorides also result in the fluxing of the nickel film. The presence of volatile or molten chlorides may be responsible for the formation of cracking and blistering of the nickel film (Figure5).

In high temperature environment, the electroless nickel film formed on iron may react with chlorine. The outer layer of the nickel film was depleted and the film thickness decreased gradually.

4 CONCLUSIONS

1. The corrosion resistance of iron against chlorination improved due to electroless nickel film formed on its surface.
2. The kinetics process of electroless nickel deposition specimens had several features. At the initial corrosion stage, the specimen showed an increase in weight according to a parabolic rate law. With increasing exposure time, the weight change of the specimens shows no change or a little decrease. After long time exposure, the protective nickel film formed on iron was completely destroyed, and the specimen showed a linear decrease in weight.
3. The electroless nickel film formed on iron was destroyed by two mechanisms: (1) mechanical damage to film by volatile ferritic chloride and (2) nickel depletion by chlorination.

References

1.   P. Elliott, A. A.Ansari, R. Prescott. M. F.Rothman. *Corrosion*, 1988, **44**, 544–554.
2.   S. Y. Lee, M. J. McNallan, *J. Electrochem. Soc.*, 1990, **137**, 472–479.
3.   M. J. Maloney, M. J.McNallan, *Metall. Trans.*, 1985, **16B**, 751–761
4.   F. H. Stott, R. Prescott and P. Elliott, *Werkst. Korros.*, 1988, **39**, 401–405.
5.   G. Y. Lai, *J. Metals*, July 1985, 14–19.
6.   C. M. Schillmoler. *Chem Eng.*, March 1980, 161–169.
7.   P. Hancock, Symposium on Halide Chemistry, Electrochem Soc., Vol. I, 1978, 645–653.
8.   P. L. Daniel and R. A. Rapp,'Advance in Corrosion Science and Technology', Plenum Press Vol.5. 1976, p. 55–172.
9.   M. H. Brown, W. B. Delong and J. R. Auld, *Industral and Engineering Chemistry*, 1947, **39**, 839–843.
10. A. S. Kim and M. J. McNallan, *Corrosion*, 1990, **46**, 746–755.
11. K. Hauffe and J. Hinrichs, *Werkst. Korros*, 1970, **21**, 954–963.
12. M. J. McNallan, Y. Y. Lee and Y. W. Chang, *J. Electochem Soc.*, 1991, **138**, 3692–3696.
13. J. W. Dini and P. R. Coronado, *Plating.*, April 1967, 385–390.
14. O. Kubaschewski and C. B. Alcock, 'Metallurgical Thermochemistry', 5th Edition, Pergamon Press. 1977, p. 78–384.

1.1.7
The Effect of Coating Thickness on the Thermal Conductivity of CVD and PVD Coatings

K. J. Lawson[1], J. R. Nicholls[1] and D. S. Rickerby[2]

[1]CRANFIELD UNIVERSITY, CRANFIELD, BEDFORD, UK

[2] ROLLS-ROYCE PLC, DERBY, UK

1 INTRODUCTION

The thermal conductivity of both physical vapour deposited zirconia and chemical vapour deposited coatings of synthetic diamond are of particular interest. The former material is used as a thermal barrier coating on gas turbine blades. While for the latter, it is an advantage for it to have higher thermal conductivity for its application in the electronics industry.

Although PVD and CVD operate under quite different pressure regimes (PVD from 10^{-7} torr to 10^{-2} torr, CVD up to 100 torr) they ar both condensation processes with the nucleation and growth processes being dependent on substrate surface energy, available nucleation sites and the atomic structure of the coating as a bulk material.

1.1 Growth of PVD and CVD Coatings

During the initial stages of growth the microcrystallites being formed will have a variety of orientations, but as film growth progresses, these regions of condensation will act as preferential nucleation sites from which will grow columns of textured material[1-4] as illustrated schematically in Figure 1. As growth continues, some columns will overshadow others and prevent their continuing formation. The early growth micro-columns develop into macro-columns which are thought to be comprised of preferentially oriented micro-columns; X-ray diffraction work has indicated this. Samples, coated using electron beam evaporation with zirconia 8wt% yttria to about 200 microns thick were examined. The metal substrates were etched off and X-ray diffraction carried out from the top and from the bottom (interface side) of the coatings. The information from the X-radiation in this case is derived principally from the first 50 microns of the material and it clearly shows that fixed textured initial growth occurs whereas at the outer part of the coating there is a dominant texture. It is noticeable that both the monoclinic and tetragonal phases are significantly present at the start of growth, whereas the monoclinic becomes dominant at the later stages of growth.

For PVD coatings Thornton's model[3,4], illustrated in Figure 2, shows the range of morphologies which reflect growth patterns, related to the substrate temperature and melting point of the condensing material (in Kelvin) and the system operating pressure. Although an idealistic model, it does indicate how the deposition parameters affect the growth of the coating. Mathematical modelling of growth and structural development through a coating also indicates the change in columnar structure and number of grain boundaries that might

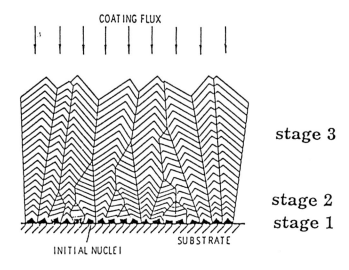

Figure 1 *Development of texture in PVD films*

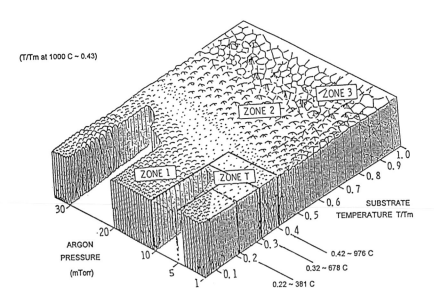

Figure 2 *Zone model applied to the deposition of yttria stabilised zirconia thermal barrier coating at 1000°C (derived from the original zone model of Thornton[3,4])*

occur[5] with coating thickness. This is in accord with the competitive growth process during which columns start growing, some become dominant and others are cut off, by these leading to a preferential growth process.

Thus the variation of thermal conductivity with thickness must reflect this competitive growth process and the associated change in morphology with coating thickness.

2 MEASUREMENT OF COATING THERMAL CONDUCTIVITY

The method used for measuring the thermal properties of the material was a laser pulse technique[6]. This is a transient method from which values of thermal diffusivity can be acquired and hence the thermal conductivity calculated.

For each zirconia-coated sample, a 1cm square was cut and the substrate thickness reduced to about 0.5mm by polishing. These test pieces were then entirely coated in colloidal graphite in order that they would be opaque to the radiation. A 0.7ms heat pulse was applied to the metal substrate from a neodymium glass laser. The temperature rise of the ceramic coating surface was monitored with time using a liquid nitrogen-cooled indium antimonide detector. A 2mm diameter spot was viewed at the sample centre. The time taken for the near sample surface to achieve half its maximum temperature rise, $t_{1/2}$ is a value used to denote the time for the heat pulse to pass through the sample. When t, and the coating thickness are known, then the thermal diffusivity can be calculated and from this, using $K = h\rho C$ (where h is the thermal diffusivity, C is the specific heat and ρ is the density) the thermal conductivity (K) can be calculated. Table 1 lists the measured thermal conductivities as a function of coating thickness determined in this study.

Graebner and coworkers[7,8] have applied a similar technique in their investigation of the thermal conductivity of CVD diamond films, but due to the high conductivity a shorter pulse time (8ns) was used. Table 2 summarises the results of Graebner et. al.[8].

3 RESULTS AND DISCUSSION

Work carried out as part of this study on electron beam, vacuum evaporated, yttria-stabilised zirconia thermal barrier coatings has shown that thermal conductivity properties are related to coating thickness (see Table 1). This trend is similar to that reported on CVD diamond

Table 1 *Thermal conductivity (at 500°C) of ZrO_2-8wt%Y_2O_3thermal barrier coatings as a function of thickness*

Coating thickness (μm)	Thermal conductivity (W/mK)
80	0.95
110	1.21
145	1.26
190	1.34
225	1.54
275	1.72
375	1.78

Table 2 *Thermal conductivity of synthetic diamond films as a function of thickness[8]*

Sample thickness (μm)	Thermal conductivity (W/mK)
30	1030
70	1450
185	1860
410	2100

films[8] both on planar silicon substrates and for fibres. Table 2 summarises the thickness dependence of CVD diamond measured on planar substrates. For both examples, this dependence seems to be related to the variation in morphology of the coating which changes through its thickness. If the morphology of the two types of coating is examined, they can be seen to have a common columnar structure as can be seen by comparing Figures 3 and 4. When the number of column boundaries through the thickness of the coating are measured (in the plane of the coating) for both systems they are very similar, see Figure 5, with the average column width increasing from 3μm at 25μm thick to 20μm at 375μm thick.

Figure 6 illustrates that the thermal conductivity of both the CVD diamond and PVD zirconia thermal barrier coating varies with thickness in a similar manner. This suggests that the coating growth morphology plays a significant rôle in determining the thermal conductivity. Texture and structure of the coating could also be important.

X-ray diffraction studies (Figure 7) have indicated changes in structure through the zirconia coatings, but thermal conductivity measurements (Figure 8) indicate no change in the coatings thermal conductivity being due to structural texturing. This strongly suggests that the growth morphology and the number of macro-column boundaries are responsible for the change in thermal conductivity of the coating with thickness.

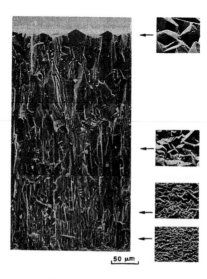

50 μm

Figure 3 *Morphology of a CVD diamond film (after Graebner, 1992[8])*

Figure 4 *Morphology of an EB-PVD zirconia thermal barrier coating, deposited at 1000°C*

Graebner and co-workers[8] measured the drop-off, in thermal conductivity, for CVD diamond (heat flow measured through the coating) and this follows a similar pattern to that measured for thermal barrier coatings as can be seen in Figure 6. Figures 4 and 3 show the morphology of a TBC as compared with the CVD schematic from Graebner's work, both have similar morphologies. TEM work carried out on the CVD coatings[9] established a concentration of defects at grain boundaries which could lead to phonon scattering and hence the reduction in conductivity with reduction in coating thickness. Phonons interact with imperfections such as dislocations, vacancies, grain boundaries, other phonons and atoms of differing masses hence if they occur particularly at column boundaries then this will lead to lower thermal conductivity particularly in the thinner coatings.

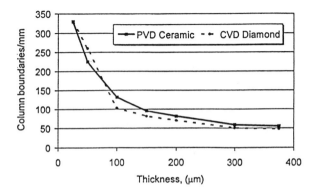

Figure 5 *Competitive growth in PVD and CVD coatings*

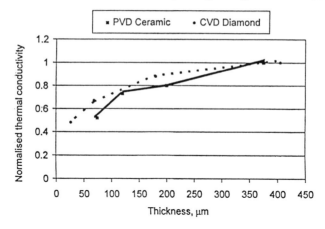

Figure 6 *Thickness dependence of thermal conductivity for PVD and CVD coatings*

3.1 A Two-Layer Model for Thermal Conductivity in Vapour Deposited Coatings

If defects at column boundaries account for the drop in thermal conductivity then it should be possible to model this behaviour and ultimately predict the thermal conductivity of the coatings.

Measurements carried out on the TBC systems showed a particular drop-off in conductivity at about 100μm, a result similar to the CVD coatings. This 100 mm region has a rapidly increasing number of column boundaries.

Figure 9 shows a schematic of the column boundary structure with two values of thermal conductivity being allocated to the coating system. The first 100μm region has a value K_1 the remainder a value K_2. The choice of the 100μm position can be justified when the column boundary density is plotted against thickness. As can be seen from Figure 5, there is an initially high boundary count which drops towards the 100μm thickness value followed by a levelling off in the boundary density value for thicker coatings. Most of the PVD coatings examined

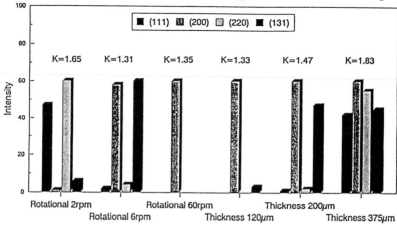

Figure 7 *Variation in texture with thickness of an EB-PVD zirconia thermal barrier coating, deposited at 1000°C*

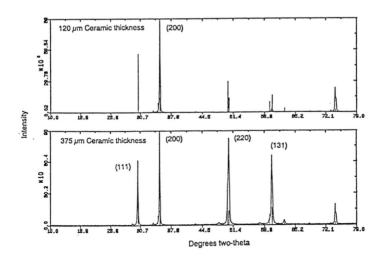

Figure 8 *Thermal conductivities of EB-PVD zirconia thermal barrier coatings as a function of texture*

show an initial confused structural growth followed by a dominating texture which persists through the main part of the coating above between 50 and 100 microns. Even the coatings which have a persistent mixed texture still follow a pattern of conductivity value relating to thickness which was the same as those coatings with a dominant outer texture. Figure 6 shows the normalised thermal conductivity curves for both PVD and CVD coatings against thickness and emphasises the similar pattern of drop-off in conductivity with thickness for both systems. As the only significant similarity between the two materials is their morphology then this again strengthens the case for the thermal conductivity change with thickness being due to number of column boundaries intersected.

Having formulated a two-layer model, it was decided to evaluate the thermal conductivity using a rule of mixtures, with the measured thermal conductivity being a weighted average of the two component parts.

Thus, for coating thicknesses (x) less than x_1

$$K = K_1 \qquad\qquad \{1\}$$

while for coating thicknesses (x) greater than x_1

$$K = \frac{xK_1}{x} + \frac{(x - x_1)K_2}{x} \qquad\qquad \{2\}$$

where $K_1 = 1.0\text{W/mK}$, $x_1 = 100\mu\text{m}$ and $K_2 = 2.15\text{ W/mK}$.

Using these constants and equation $\{2\}$ an expected thermal conductivity thickness dependence can be derived for coating thicknesses greater than $100\mu\text{m}$. The predicted thermal

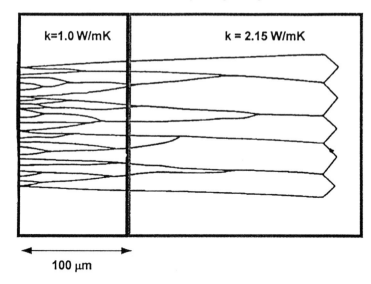

Figure 9 *Two-zone model for the thermal conductivity of a EB-PVD zirconia thermal barrier coating*

conductivity has been plotted against the measured thermal conductivity for various thicknesses of ceramic in Figure 10. The linear dependence shown in Figure 10, demonstrates the good fit of this two-layer model. The 95% confidence limits of this prediction are ± 0. 1 W/mK. Also, by taking mean values of thermal conductivity and plotting them against these confidence bands, Figure 11 shows how the mean data files between the predicted upper and lower boundaries over the thickness range 100–400µm, further demonstrating the ability of this two-layer model to predict the thermal conductivity of vapour depositing coatings as a function of thickness.

4 CONCLUSIONS

The PVD zirconia thermal conductivity can be modelled using a two-layer system with an accuracy of about 0. 1W/mK. The increase in thermal resistance close to the interface region is the result of the greater number of column boundaries. Such behaviour has also been reported as the result of competitive growth processes in CVD diamond films. Texture and crystal structure have a minor effect on thermal conductivity for a specific material compared with morphology.

Acknowledgements

The authors wish to thank Rolls-Royce for funding this research and partners in the joint development programme on Advanced TBC Development (Rolls-Royce, CUK Ltd, the Defence Research Agency and AEA Technology) for useful discussions throughout the course of this work and permission to publish this paper.

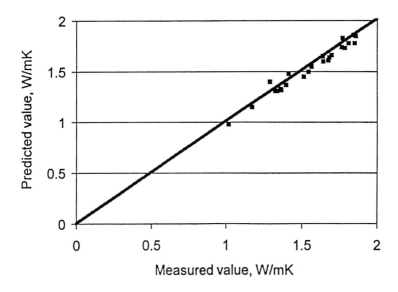

Figure 10 *Comparison between predicted and measured thermal conductivities for EB-PVD zirconia thermal barrier coatings*

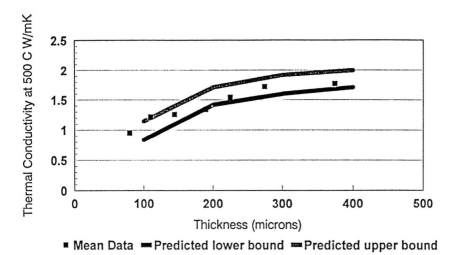

Figure 11 *Variation of mean thermal conductivity with thickness, plus predicted tolerance bands*

Refererences

1. B.A. Movchan and A.V. Demchishin, *Phys. Met. Metall*, (1969), **28**, 83.
2. C.R.M Grovenor, H.T.G. Hentzell and D. A. Smith, *Acta. Met.* (1984), **32**, 773.
3. J. A. Thornton, *J. Vac. Sci. Tech.*, (1974) **11**, 666.
4. J. A. Thornton, *Ann. Rev. Mater. Sci.* (1977), **7**, 239.
5. A. Mazor, D. J. Srolovitz, P. S. Hagan and B. G. Bukiet *SPIE*, 'Modelling of Optical Thin Films', (1987), **821**, 88.
6. British Standard, BS7134: Section 4.2 'Method for the Determination of Thermal Diffusivity by the Laser Flash (or Heat Pulse) Method', 1990.
7. J. E. Graebner, S. Jin, G. W. Kammlott, B. Bacon, L. Seibles and W. Banholzer, *J. Appl. Phys.*, (1992), **11**, 5353.
8. J. E. Graebner, S. Jin, G. W. Kammlott, J. A. Herb and C. F. Gardinier, *Nature*, (1992), **356**, 401.
9. A. V. Hetherington, C. J. H. Wort and P. Southworth, *J. Mater. Res.*, (1990), **5**, 1591.

Section 1.2 Aqueous Corrosion

1.2.1

Electrochemical Studies and Characterisation of Electrodeposited Zinc-Chromium Alloy Coatings

M. R. El-Sharif[1], C. U. Chisholm[1], Y. J. Sut[2], L. Feng[3]

[1]DEPARTMENT OF ENERGY AND ENVIRONMENTAL TECHNOLOGY, GLASGOW CALEDONIAN UNIVERSITY, CITY CAMPUS, 70 COWCADDENS ROAD, GLASGOW, UK
[2]CHARTERED SEMICONDUCTOR MANUFACTURING LTD, SINGAPORE
[3]SHANGHAI RESEARCH INSTITUTE OF IRON AND STEEL, 1001 TAI HE ROAD, SHANGHAI, PR CHINA

1 INTRODUCTION

In recent years many electrodeposited zinc alloy coatings have been studied in an attempt to achieve greater corrosion resistance with smaller coating thickness [1,2]. To provide sacrificial type of corrosion protection for steel substrates is the main concern of these zinc alloy coatings. Electrodeposited zinc-chromium alloy has been investigated because of its ability to self-passivate, and exhibit and always maintain a less noble potential than steel and thus keep its galvanic protection over the steel substrate. The alloy coatings have been deposited from an environmentally more acceptable chromium (III)-based electrolyte system, although the deposition was encountered with difficulties due to the complex chemical behaviours of chromium (III) species in aqueous solutions. The studies of zinc-chromium alloy deposition have been previously published[3]. Another important feature of zinc-chromium alloy is that corrosion resistant conversion coating, other than a chromate-formed conversion coating, can be generated on the surface of the alloy coating by surface oxidation. Chromate conversion coatings are highly effective in protecting zinc and its alloys against the formation of an initial corrosion product known as the 'white rust', but chromating treatments suffer from highly toxic, carcinogenic and environmentally undesirable chromium (VI) solutions[4]. The formation, on zinc-chromium alloy, of a conversion film similar to a chromate one was achieved by surface oxidation using a simple immersion process, thus eliminating the use of chromium (VI) treatment solutions. In this paper electrochemical assessments of zinc-chromium alloy with and without a conversion film are described. These assessments were made by methods of potentiodynamic polarisation and impedance analysis. This paper also includes the characterisation of the alternative conversion film by using X-ray photoelectron spectroscopy (XPS), Auger electron spectroscopy (AES) and structural analysis of the alloy deposits both before and after heat treatments by using X-ray diffraction (XRD) technique.

2 EXPERIMENTAL

Electrodeposited Zn-Cr alloy coatings were obtained from the bath shown in Table 1. Mild steel cathodes were used as substrates for deposition of the alloy. The cathodes were polished with silicon carbide paper grade 500, degreased in acetone, and coated with the lacquer to leave a working area of 2.5×2.5 cm. They were then treated cathodically in a solution containing

50–70 g/1 alkaline cleaning agent with a current density of 50 mA/cm² at 800°C for one minute, followed by a rinse in deionised water. High density graphite was used as an anode. Electrodeposition was carried out under potentiostatic mode, with circulating electrolyte to ensure a sustained deposition. Cathode vertical movement was employed to minimise the destructive effect of gas bubbles thus improving the quality of the deposits. The composition of the alloys was determined by X-ray fluorescence in a JXA–50A electron microscope attached with a Link electron microprobe analysis unit.

In the formation of conversion films electrodeposited alloys were pre-treated in a 3% (v/v) nitric acid solution for 10 seconds and then rinsed in deionised water. This was followed by the immersion of the coatings in a solution containing an oxidant and other agents. The coatings were rinsed again in deionised water, and dried in hot air or idled for a certain period of time. Table 2 gives compositions of several solutions which were used in the study of conversion coating formation.

Potentiodynamic polarisation measurements were carried out for the alloy coatings in a non-deaerated 1.0 M NaCl solution of pH 6.0 and at 300°C, using an EG & G-352 corrosion measurement system. The anodic and cathodic polarisation curves in the Tafel or linear region were obtained with a typical potentiodynamic scan rate of 10 mV/min. The corrosion currents, I_{corr}, can be calculated from Tafel constants (β_a and β_c) which were obtained from Tafel plots, and from the Polarisation Resistance, R_p, which was determined from the slope at the origin of the polarisation curve, using the Stern–Geary equation[5].

$$I_{corr} = \frac{\beta_a \beta}{2.303 \times R_p (\beta_a + \beta_c)} \quad \{1\}$$

In the present case where dissolved oxygen exists in the solution it was assumed that the cathodic reaction was diffusion controlled and $\beta_c \gg \beta_a$. Thus the corrosion currents can be calculated form a simple form of equation {1} as:

$$I_{corr} = \frac{\beta_a}{2.303 R_p} \quad \{2\}$$

Table 1 *Composition of bath used in Zn-Cr alloy deposition*

Component	Concentration (g/l)
$ZnSO_4.7H_2O$	57
$CrCl_3.6H_2O$	215
Urea	240
Secondary complexant	38
H_3BO_3	9
NH_4Cl	27
NaCl	29
Bulk pH	2.0–3.0
Temperature (°C)	20–25

AC impedance measurements were also conducted to give more information about the corrosion properties of the alloy coatings. The equipment used for the impedance measurements was a Sycopel TFA2000A AC impedance analyser. The range of frequency used was from 100 kHz to 0.01 Hz, and the amplitude of the measuring voltage was 5 mV. The measurements were performed at 200°C in a non-deaerated 0.6 M NaCl solution of pH 6.0.

A Perkin–Elmer PHI 550 ESCA/SAM electron spectrometer with an Mg Kα X-ray (1253.6 eV) anode was used for XPS and AES analysis of the conversion film formed on the coatings. Survey and high-resolution XPS spectra were obtained with the energy analyser operating in a constant transmission energy mode at pass energies of 100 and 50eV. the pressure in the analyser chamber was maintained at less than 10^{-7} Pa during analysis. The voltage and current of the electron beam for the AES depth profile were 3keV and 1A respectively. An argon ion gun with voltage 3kV and emission current of 100 A/cm^2 was used for depth profiling studies, the spotting area was 1 x 1 mm. Binding energies were corrected for charging effects by reference to Cls peak (284.6eV) and their estimated error was 10.1 eV.

Microstructure changes of the alloy deposits before and after heat treatments were investigated by continuous scanning X-ray diffraction method with a Phillip PW1840 diffractometer. The radiation used was a monochromatic CuKα. The heat treatments were conducted at 120, 300 and 500°C respectively in argon atmosphere for one hour.

3 RESULTS AND DISCUSSION

3.1 Electrodeposition of Zinc-Chromium Alloy

Table 1 shows the bath used to deposit Zn-Cr alloy coatings. It has been found that the deposition of Zn-Cr alloy reveals characteristics of anomalous co-deposition with zinc as a less noble metal being preferentially deposited. Concentrations of zinc and chromium in electrolytes are both critical for the quality and chromium content of Zn-Cr deposits.

From a low zinc bath (< 0.1 mol/l zinc sulphate) thin deposits containing up to 40%Cr could be obtained, but thick deposits with a moderate level of chromium content were obtained only when the zinc concentration was maintained at near 0.2 mol/l and chromium concentration above 0.8 mol/l. Current efficiency was found to be up to 45% in the optimum range of cathode potential. Vertical cathode movement was found to be effective in minimising the negative influence of the hydrogen gas bubbles on deposit quality. The results of the bath operation parameters are previously reported[6].

Table 2 *Solutions used in formation of conversion coating*

Composition	I	II	III	IV
$Na_2Cr_2O_7.2H_2O$ (g/l)	200			
$KMnO_4$ (g/l)		5	10	
$Na_2S_2O_8$ (g/l)				5
HNO_3 (ml/l)	5	10	10	10
H_2SO_4 (ml/l)	10	20	20	20

3.2 Studies of Formation of Conversion Coatings

3.2.1 Mechanism Formation of Conversion Film on Zinc Alloys. The solutions used for the formation of chromium-based conversion film on zinc alloys were highly acidic (pH = 1.5–1.8) containing oxidant such as permanganate, film-forming activators, and organic acids. In the case of zinc-chromium, as the alloy is immersed into such a solution zinc from the alloy is preferentially dissolved and reacts with hydrogen ions in the metal-solution interface.

$$Zn + 2H^+ \longrightarrow Zn^{2+} + H_2 \qquad \{3\}$$

Chromium content near the alloy surface is enriched and activated and its dissolution takes place after being oxidised by an oxidant such as permanganate, represented simplistically by

$$5\,Cr + 3\,MnO_4^- + 24\,H^+ \longrightarrow 5\,Cr^{3+} + 3Mn^{2+} + 12H_2O \qquad \{4\}$$

A portion of chromium (III) is expected to be further oxidised into chromium (VI)

$$10Cr^{3+} + 6MnO_4^- + 11H_2O \longrightarrow 5\,Cr_2O_7^{\,2-} + 6Mn^{2+} + 22H^+ \qquad \{5\}$$

As reactions {3} and {4} can be regarded as predominating reactions the hydrogen ions within the metal-solution interface are consumed very quickly and this results in a rapid increase of pH in the interface. When pH reaches a critical value a complex chrome gel containing chromium (III) and chromium (VI) oxo/hyrdoxo species as well as zinc hydroxide and zinc chromate species is formed and precipitated on the surface of the alloy forming the conversion

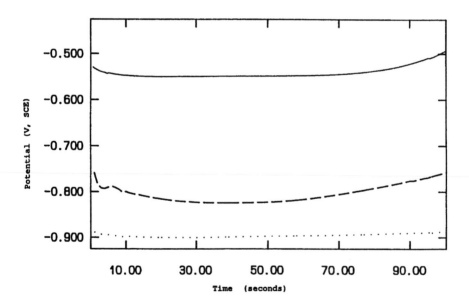

Figure 1 *(a) Potential-time curve for zinc in chromate (——), Zn-Cr (4% Cr) in 5g/l KMnO$_4$ (------), and Zn-Cr (4% Cr) in 5g/l Na$_2$S$_2$O$_8$ (.........) at 20°C*

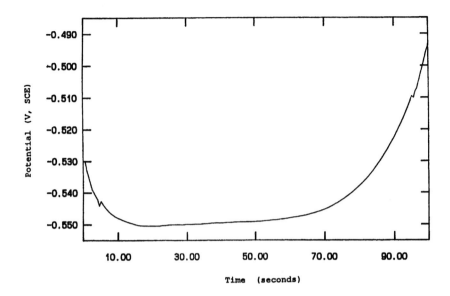

Figure 1 *(b) Potential-time curves for zinc in chromate at pH 1.8 and 20ºC*

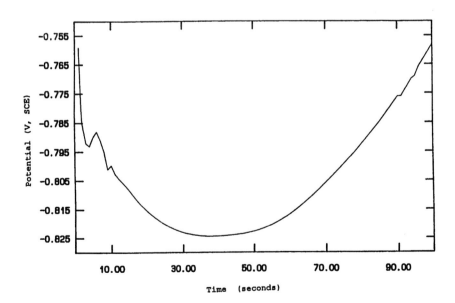

Figure 1 *(c) Potential-time curves for Zn-Cr (4% Cr) in 5g/l KMnO₄ at pH 1.8 and 20º*

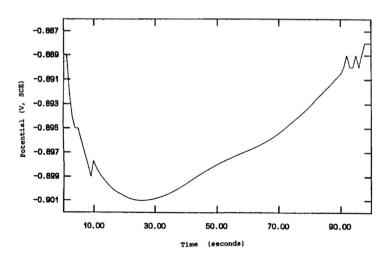

Figure 1 *(d) Potential–time curves for Zn-Cr (4% Cr) in 5g/l Na₂S₂O₈ at pH 1.8 and 20°C*

film. The above proposed composition of the conversion film is supported by results obtained from AES and XPS.

3.2.2 Effect of Oxidising Agent and Concentration. Figure 1 (a) shows the potential-time curves obtained in the formation of conversion coating for Zn–Cr alloys of 4% Cr in solutions at 20°C with different oxidising agents, which is also compared with that obtained from chromate treatment of zinc. The solutions used in the formation of conversion films are listed in Table 2. Details of each curve in Figure 1 (a) can be seen when plotted in their own corresponding

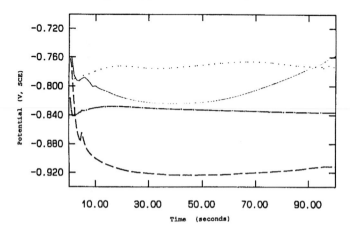

Figure 2 *Potential-time curves for Zn-Cr (4% Cr) in 5g/l KMnO₄ (_____), Zn-Cr (4% Cr) in 10g/l KMnO₄ (- - - - - - -), Zn-Cr in 5g/l KMnO₄ (............) and Zn-Cr (4% Cr) in 10g/l KMnO₄ (-.-.-.-.-.-.-) at 20°C*

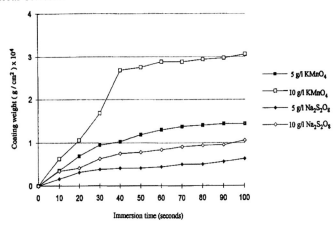

Figure 3 *Effect of oxidizing agent and its concentration in the conversion treatments on weight gain of conversion coating with Zn-Cr at 20°C*

potential range in Figure 1 (b)–(d) respectively. It is noted that all curves show an initial decrease (more electronegative) in potential, indicating the metal surfaces were effectively activated. The activators in chromate solution to zinc are well known as chloride and sulphate[7], which prevent an initial passivation from occurring by facilitating the dissolution of zinc. Some of the present solutions contained sulphate ions which obviously revealed activation in the absence of chromate. Iridescent conversion films were obtained from these solutions. It is also noted that each curve has a major peak at about 10 s and the reason is under investigation.

The potential started to increase at different points of time depending on the solution

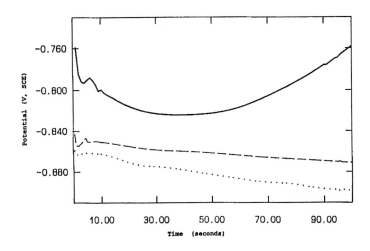

Figure 4 *Potential–time curves for Zn-Cr (4% Cr) in 5g/l KMnO$_4$ at various temperatures: 20°C (———), 40°C (--------), 50°C (...........)*

period of immersion time such as 30 s produced thinner film from persulphate than other solutions, and this may be associated with the shorter period when the surface of an alloy was in an activated state in the persulphate. Also from Figure 1(a) the variations of potential in chromate and permanganate solutions within 100 s were similar, 55 mV and 65 mV(SCE) respectively, while the potential variation in persulfate was 13 mV (SCE).

The effect of permanganate concentration on the potential-time of Zn-Cr (4%Cr) can be seen in Figure 2. Unlike the 5 g/1 $KMnO_4$ solution, the potential in 10 g/1 $KMnO_4$ decreased more rapidly at the beginning and increased very slowly thereafter and its overall potential was shifted towards more negative values. Thicker conversion films were obtained from this more concentrated permanganate solution. The effect of oxidising agents and their concentration on weight gain of conversion film was studied, on Zn–Cr alloys of 4%Cr in permanganate and persulfate solutions (Figure 3). A general rise of film weight gain with concentration of oxidising agent was found, markedly in 10 g/1 $KMnO_4$ in the initial 40 sec. The increase of persulfate concentration had less effect on the weight gain of the films than that of permanganate concentration.

3.2.3 Effects of Chromium Content of Alloy and Immersion Time. The chromium content of the alloy is significantly important to the chromium content and state in the conversion coating as it is the only source providing chromium species needed in the conversion coating. A number of Zn-Cr alloys containing chromium between 0% and 10% were used to form permanganate conversion coating and the visual surface appearance is described in Table 3. When immersion times were less than 30 seconds light yellow conversion films were obtained on the alloys of 2%Cr while iridescent films were obtained from the alloys of 6% and 10% Cr. Although film thickness was one of the factors influencing the film colour there was a correlation between the chromium percentage of alloy and the distribution ratio of green-blue to orange-red area of the film surface under the optical microscope. The higher the chromium content of the alloys the higher the ratio of the orange-red colour was found, provided the solutions and operating conditions were identical, indicating a larger portion of chromium from the alloy was transformed into hexavalent chromium. Immersion times beyond 60 seconds produced thicker films with increase in micro-cracking.

The potential of Zn-Cr with 8% Cr was found to be less dependent on time of immersion despite the initial sharp decrease observed both in 5 and 10 g/1 $KMnO_4$ solutions at 20°C (Figure 2). In comparison with 4%Cr alloys, the 8%Cr resulted in more positive potentials in the solutions.

Table 3 *Surface appearance of conversion films on Zn-Cr alloys (obtained from 5 g/l $KMnO_4$ solution, pH = 1.5-1.8 at 25°C)*

Immersion time (seconds)	Chromium content of alloy		
	2%	6%	10%
10	Very light yellow	Light-iridescent	Iridescent
20	Light yellow	Irdescent	Iridescent
40	Light yellow	Iridescent	Dark-iridescent
60	Dark yellow	Dark-iridescent	Brown-iridescent
90	Yellow-brown	Brown-iridescent	Brown-iridescent

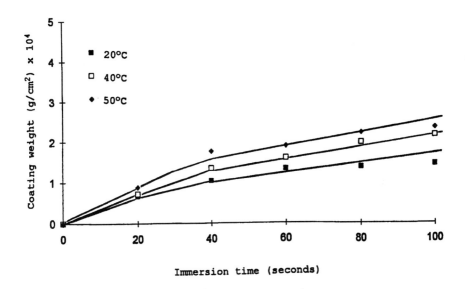

Figure 5 *Effect of temperature in the conversion treatments on weight gain of conversion coating, withZn-Cr (4%Cr) and at pH 1.8*

3.2.4 Effects of Temperature and pH. The effect of temperature on potential-time profile was also investigated (Figure 4). The increase of temperature shifted the surface potential of Zn-Cr alloys (4%Cr) towards a more negative value. At 20°C the potential reached a maximum negative value before becoming more positive, but at 40°C and 50°C the potentials decreased throughout the 100 seconds of immersion except for the first 10s, and the range of variation was found to be not too large. This longer period of surface activation is associated with the diffusion rate of active species within the metal/solution being increased with temperature. The effect of temperature on weight gain of the above alloys is shown in Figure 5, where the rise in temperature caused increase in weight although the difference is not significant. However, the conversion films formed at higher temperatures generally had rougher surfaces and micro-cracks being developed. The pH below 1.0 resulted in a rapid dissolution of initially formed conversion film, eventually the alloy coating itself. The increase of pH to above 2.0 generally produced dark films and increasingly less iridescent colour.

3.2.5 Ageing of the Conversion Coating. The as-formed conversion coatings were normally weak and susceptible to scrape damage. The ageing of the coatings was conducted in two ways, drying in hot air flow at 60°C and drying in air at room temperature for 24 hours. Stronger films were obtained from the former technique. The surface appearance of the films was also found somewhat different by using these two different drying techniques, with more iridescent colour from the higher temperature drying. The structure of the films and the chromium state are expected to be influenced by higher temperature drying, as the molecular weight of the polymeric species can be changed and the conversion of hydroxo bridging to oxo bridging can be speeded up by the increase of temperature.

3.3 Electrochemical Investigation of Conversion Coatings

3.3.1 Potentiodynamic Polarisation Measurements. The potentiodynamic polarisation measurements were conducted for Zn-Cr alloys before and after conversion treatments, to study the effect of the conversion film on electrochemical properties. The results are shown in Table 4, in which data for zinc before and after chromate conversion treatments are also included for comparison. The chromated zinc samples were prepared by immersion of electrodeposited zinc coatings into *solution I* indicated in Table 2. It can be seen from the trend of corrosion potential variation of non-passivated Zn-Cr alloys, that adding chromium to zinc coating shifted the corrosion potential towards the noble side compared with that of Zn in proportion to the content of chromium. Zn-6%Cr showed a corrosion potential of approximately −1000 to −1030 mV(SCE), far less noble than Zn-Fe and Zn-Ni deposits.

The results presented in Table 5 show the effect of immersion time and the concentration of $KMnO_4$ in the conversion treatments. The corrosion potential of the coatings obtained from 10 g/l $KMnO_4$ were generally more positive than those obtained from 5 g/l $KMnO_4$ and their corrosion currents were found to be much higher. The activation/passivation behaviours of Zn-Cr (4%Cr) in conversion treatment can be seen in Figure 2, where, in contrast to 5 g/l $KMnO_4$, the activation of the surface in 10 g/l $KMnO_4$ was carried out at much more negative potentials and the passivation was not clearly seen as the potential remained much the same after the initial 20s. This indicates a severe dissolution and a possibility of losing chromium content and other species needed in the conversion films into the bulk solution even when the immersion time was properly controlled, resulting in poor quality of the coatings. The immersion time of 40s was found to be optimum in 5 g/l $KMnO_4$ for producing conversion films that have lower corrosion rate. This suggested that to obtain the best conversion films the immersion should be completed at the time after the activation step was completed plus a short period when the surface was in the activated state and the dissolution-oxidation-precipitation process still continued (Figure 1c). The passivation can be then completed during the removal and the rinsing of the samples. However, this pattern did not fit the case

Table 4 *Effect of formation of conversion film on corrosion properties*

Non-passivated Zn-Cr alloy			Passivated Zn-Cr alloy		
Cr% of alloy	E_{corr} (V, SCE)	I_{corr} ($\mu A/cm^2$)	Cr% of Alloy	E_{corr} (V, SCE)	I_{corr} ($\mu A/cm^2$)
0*	−1.06	23.1	0**	− 0.96	5.2
2.2	−1.05	25.6	2	− 0.97	9.4
4.3	−1.04	20.7	4	− 0.93	3.1
6.1	−1.03	17.4	6	− 0.92	2.8
10.2†	−1.02	26.1	8	− 0.90	7.4

*Electrodeposited zinc coating **Electrodeposited zinc with chromate treatments
†Thin coatings (2–3 μm)

Table 5 *Effect of KMnO$_4$ concentration and immersion time on corrosion properties of conversion film (Zn-4%Cr, at 20°C)*

KMnO$_4$ concentration (g/l)	Immersion time (sec)	E$_{corr}$ (V, SCE)	I$_{corr}$ (mA/cm^2)
5	20	− 0.95	5.5
5	40	− 0.93	3.1
5	60	− 0.97	6.8
10	20	− 0.92	8.0
10	40	− 0.94	17.7
10	60	− 0.96	17.6

in 10 g/1 KMnO$_4$, in which immersion time on 40s or beyond resulted in much higher corrosion rate[5].

Table 6 shows the effects of chromium content in alloys and the operating temperature of conversion treatments on corrosion parameters. It is found that as the chromium content of the alloys increased, the corrosion potential shifted towards more positive direction and the corrosion currents decreased, indicating more chromium species were contained in the conversion coatings. Those coatings obtained at higher temperature conversion treatment revealed poorer corrosion resistance but no significant change in corrosion potential was found as a result of change in temperature. Table 7 shows the effect of different solutions used in the conversion treatments. There was no significant difference in corrosion potential and corrosion current between using 5 g/1 KMnO$_4$ and 5g/1 Na$_2$S$_2$O$_8$.

Temperature such as 50°C yielded films with poorer corrosion resistance. The pH of the treatment solution is also an important factor. In Table 8, it can be seen that higher corrosion currents were measured for conversion films obtained either at pH 1.0 or pH above 2.5, suggesting the optimum range of pH is about 1.5–2.0.

3.3.2 Impedance Measurement. Figure 6 shows impedance diagrams generated from zinc and Zn-Cr alloy coatings. It is clear that the semi-circle for zinc is completed. The semi-circle for Zn-Cr alloy is not completed even at the lowest frequency of 0.01 Hz, but it is relatively well defined. As a result of alloying chromium to zinc the capacitive loop is largely broadened. For Zn-Cr alloy the extrapolation of its impedance values gives charge transfer

Table 6 *Effect of chromium in alloy and treatment temperature on corrosion properties of conversion film (immersion time 40 sec, in 5g/l KMnO$_4$)*

Cr(%)	Temperature (°C)	E$_{corr}$ (V, SCE)	I$_{corr}$ (μmA/cm^2)
4	20	− 0.93	3.1
4	40	− 0.97	8.3
4	50	− 0.94	12.0
6	20	− 0.92	2.8
6	40	− 0.92	6.3
6	50	− 0.93	12.4

Table 7 *Effect of type of treatment solution on corrosion properties of conversion (Zn-4%Cr, immersion time 40 sec, in 5g/l KMnO$_4$, 20°C)*

Treatment solution	Solution concentration	E_{corr} (V, SCE)	I_{corr} (mA/cm^2)
KMnO$_4$	5	− 0.93	3.1
Na$_2$S$_2$O$_8$	5	− 0.99	4.2

resistance of 2000 Ωcm^2, which is compared with the one for zinc of 120 Ωcm^2.

In Figure 7, impedance diagrams for chromated zinc and permanganated Zn-Cr alloy are given. Although semi-circles are not clearly demonstrated in three curves, half of a semi-circle at high frequencies and a tail at low frequencies can be seen in all cases. For commercial chromated zinc and permanganated Zn-Cr alloy the tails represent straight lines, while for laboratory prepared chromated zinc the tail levels off at even lower frequencies. It can be seen that the curves for laboratory prepared chromated zinc and permanganated Zn-Cr alloy are very close.

3.4 Structural Studies

3.4.1 X-ray Diffraction. The X-ray diffraction analysis was conducted for non-passivated Zn-6%Cr alloy deposits both in the as-plated state and after heat treatment at various temperatures.

From Figure 8, for as-plated deposit copper peaks and two other sharp peaks were obtained. These two sharp peaks cannot be identified by using standard diffraction files. In previous studies, from electrolytes containing similar ingredients as in the present bath, chromium and other chromium alloy deposits gave no sharp peak while zinc deposits gave peaks related to a number of crystal planes of zinc[8,9]. These are not observed in the present pattern, so that it can be taken as an indication that homogeneous Zn-Cr solid solution was formed and the alloy is crystalline. An identical pattern was obtained for the deposit which was heat treated at 120°C. At such a temperature change the structure of the alloy deposit can be excluded. More diffractions were obtained for the deposit heat treated at 300°C. Further examination of these diffractions show that they are correlated with each other, giving a body-centred-cubic structure for the alloy. Heat treatment of the deposit at a higher temperature of 500°C yielded more diffractions, and some of the newly appearing peaks evidently belong ZnCrO$_4$. Three out of four diffractions which are believed to be related to the equilibrium Zn-Cr alloy remain.

Table 8 *Effect of pH in treatment solution on corrosion properties of conversion film (Zn-4%Cr, immersion time 40 sec, in 5g/l KMnO$_4$, 20°C)*

pH	E_{corr} (V, SCE)	I_{corr} (μA/cm^2)
1.0	− 0.97	13.6
1.8	− 0.93	3.1
2.5	− 0.94	8.4
3.5	− 0.95	9.8

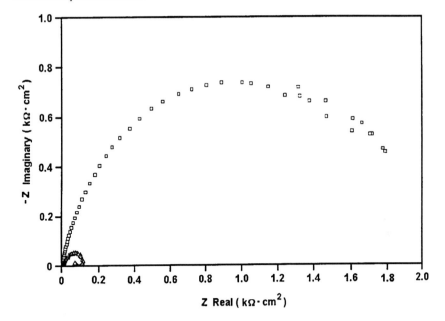

Figure 6 *Nyquist plots of impedance scans of zinc (Δ) and Zn-Cr alloy (□) immersed in aerated 0.6 MNaCl solution at 30°C*

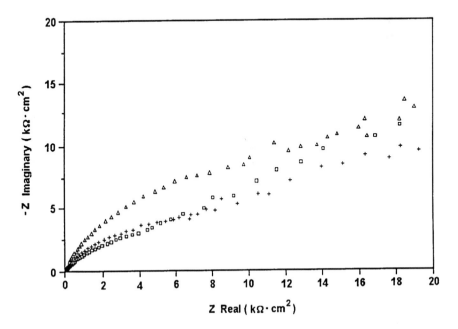

Figure 7 *Nyquist plots of impedance scans of commercial chromated zinc (Δ), laboratory prepared chromated zinc (+), and permanganated Zn-Cr alloy (□) immersed in aerated 0.6M NaCl solution at 30°C*

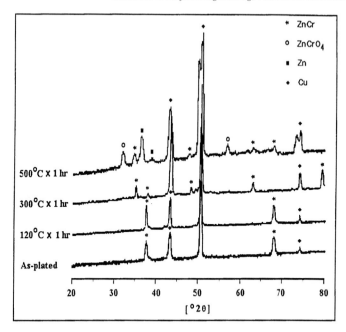

Figure 8 *The X-ray diffraction pattern of as-plated and heat treated Zn-Cr alloy coatings*

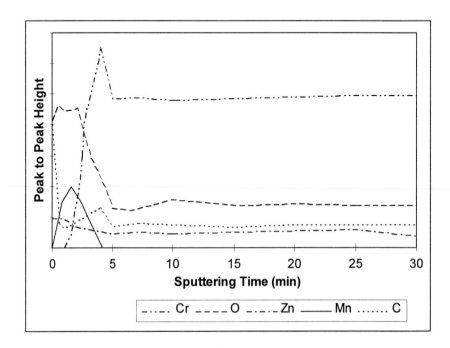

Figure 9 *Depth profile of composition determined by AES for conversion film formed on Zn–Cr coating treated in 5g/l KMnO₄*

Other peaks were not identified by standard files. It is worthwhile to note that there is no diffraction of zinc oxide, chromium oxide or chromium carbide from the pattern. At this temperature, formation of certain forms of oxide is a possibility, but it is likely to involve both zinc and chromium.

3.4.2 Auger Electron Spectroscopy. Examination of the surface of the conversion film formed on the Zn-6%Cr alloy deposit, treated in 5 g/1 $KMnO_4$ solution, by AES revealed that the film contained Cr, C, O, Zn and Mn. Figure 9 shows the distribution of the above five elements in the film as a function of sputtering time. The conversion layer was about 500Å thick and was probably a mixture of zinc oxides and chromium oxides. The surface carbon and oxygen were very high presumably due to environmental impurities. The sharp rise in chromium peak intensity between 0–5 minutes sputtering time supports the mechanism proposed in that the chromium content near the alloy surface becomes enriched when the zinc from the alloy is preferentially dissolved.

3.4.3 X-Ray Photoelectron Spectroscopy. XPS was used to understand the valence and type of bonding in the conversion film obtained. A peak corresponding to $Cr2P_{3/2}$ at 477.0 eV resulted from chromium oxides, Cr_2O_3 (576.6eV), or CrOOH (576.8ev). Oxygen in the Cr-O bond revealed a peak at 531.3eV. Accordingly, the O1s peak at 531.5ev can be assigned to oxygen in CrO_3 bonds. These values show that there are both Cr(III) and Cr(VI) in the film. The existence of Cr(III) and Cr(VI) in the film supports the mechanism proposed for the formation of conversion film on Zn-Cr alloys. The binding energy peak for Zn is 1022.4eV which can be assigned to that of ZnO $(1022.2eV)^{10}$.

The composition of the conversion film obtained in this investigation is similar to that reported in the literature. According to the ESCA investigations of chromate conversion coatings[11] chromium exists in three forms in the film. The dominant form is trivalent chromium, the remainder being metallic chromium and the hexavalent chromium.

4 CONCLUSIONS

1. Thick deposits of high quality Zn-Cr alloy with chromium content up to about 10% can be obtained from an acid zinc based chromium (III) aqueous electrolyte.
2. X-ray diffraction analysis of Zn-6% Cr alloy deposit in the as-plated state revealed that the alloy does not contain free Zn and chromium, instead it has sharp peaks which are considered to be a sort of Zn-Cr alloy phase with non-equilibrium structure. On heat treatment above 300°C the structure transformed to an equilibrium structure with the appearance of new diffraction lines belonging to $ZnCrO_4$.
3. Characteristically similar conversion film can be formed by immersion of Zn-Cr deposits into acidic permanganate ($KMnO_4$) and persulphate solutions ($Na_2S_2O_8$).
4. The quality of the conversion film is influenced particularly by treatment time, chromium content of alloys, and concentration of oxidising agents in the treatment solutions.
5. Conversion coatings obtained from persulphate are generally thinner and lighter than those obtained from permanganate. However, the electrochemical measurements show little difference in corrosion properties between them.
6. The best conversion coatings can be obtained at 5-10s after the activation process is completed in a specific solution, but this time-controlling method is not applied to the case when severe dissolution takes place during the conversion treatment.

7. A combination of operating conditions which results in long or virtually no activation period is not favourable to good quality conversion coatings.
8. The XPS measurements indicate that the conversion film consists of chromium oxides (Cr (III)) and (Cr (VI)) and zinc oxides. This composition resembles the composition obtained using chromate treatment the valence and type of bonding in the conversion film obtained.

References

1.   A. P. Shears, *Trans. Inst. Metal Finish*, 1989, **67**, 67.
2.   R. R. Sizelove, *Plating and Surf. Fin.*, 1991, **78**, 26.
3.   M. R. El-Sharif, Y. J. Su, C. U. Chisholm and A. Watson, Proc.of the International Technical Conference of SUR/FIN'93, USA, 1993.
4.   P. Kodak, *Plating and Surf. Fin.*, 1989, **76**, 30.
5.   M. Stern, *Corrosion J.*, 13 Sept. 1958, 440.
6.   Y. J. Su, A Watson, M. R. El-Sharif and C. U. Chisholm, Surface Engineering, Vol. 1 'Fundamentals of Coatings' Ed by P. K. Datta and J. S. Gray, Pub. by Royal Society of Chemistry, 1993, p. 78.
7.   L. F. G. *Williams, Surface Technology*, 1976, **4**, 355.
8.   D. J. Willis and C. Hammond, *Materials Science and Technology*, 2 July 1986, 630.
9.   M. R. El-Sharif, S. Ma and C.U. Chisholm, The NACE International Annual Conference and Corrosion Show - Corrosion '95, Paper No. 595.
10. 'Practical Surface Analysis by AES and XPS' Ed. by D. Briggs and M.P. Seah Pub. by John Wiley and Son 1983, p. 498.
11. S. J. Lemperueur, J. Renard and V. Leroy, *Br. Corros J.*, 1979, **14**, 126.

1.2.2

Comparative Study of Corrosion Properties of Zinc Composites in Acidic and Alkaline Environments

D. Sneddon[1], M. R. El-Sharif[2], C. U. Chisholm[3], A. M. DeSilva[4]

[1]DEPARTMENT OF ENERGY & ENVIRONMENTAL TECHNOLOGY

[2]FACULTY OF SCIENCE & TECHNOLOGY

[3]DEPARTMENT OF ENGINEERING, GLASGOW CALEDONIAN UNIVERSITY, 70 COWCADDENS ROAD GLASGOW, G4 0BA, U.K.

1 INTRODUCTION

Coatings that display improved engineering properties such as resistance to wear and corrosion have attracted a fair amount of attention. Composite coatings fall into this category. One of the most important techniques for producing composite coatings of metallic and non-metallic constituents is electrodeposition. Many composite coatings have been developed in recent years for a variety of applications; namely: oxidation and wear resistance[1-3], dispersion strengthened coatings[4,5] and self-lubricating deposits[6]. An extensive literature review has indicated that limited efforts have been made to improve the properties of zinc MMCs[7,8]. The extensive use of zinc as a sacrificial coating within the metal finishing industry is one of the few growth areas. An example is the use of zinc coated steel for automobile body parts. A number of these panels may be employed near engine components and the conduction of heat from these components will mean the panels must be able to withstand higher temperature operating conditions while maintaining the desired properties. Increasing use of road salts to combat ice has led to a harsher acidic corrosive environment. The present paper presents the results of a new attempt to improve the corrosion properties of electrodeposited zinc composite coatings in both acid and alkali conditions to simulate road conditions and alkaline marine environment. Particles of Al_2O_3, Cr, SiO_2 and TiO_2 were dispersed within an acetate/chloride-based electrolyte. The corrosion behaviour and morphology of the resulting MMCs were investigated as-plated and after heat treatment of 100°C, 150°C and 200°C in an open furnace and under inert (argon) conditions. These samples were compared with that of zinc and zinc alloys.

2 EXPERIMENTAL DETAILS

The electrolytes employed in this study are given in Table 1. The deposits in this work were zinc and zinc composites (~25µm thick) obtained on mild steel substrates.

The microstructural characterization was performed using a scanning electron microscope (SEM) and surface microanalysis was determined using energy dispersive X-ray spectroscopy (EDXS). Checks on coating composition were made by stripping, in a fresh solution of hydrochloric or nitric acid, followed by solution analysis using atomic absorption spectroscopy. In order to evaluate the heat treatment effects on the coatings' corrosion resistance, some

Table 1 *Electrolyte & Particle data*

Chemical	Molarity (M)	Particle type	Mean size (μm)	Density (g/cm^3)
ZnCl$_2$	0.25	Al$_2$O$_3$	1	3.96
Na(CH$_3$COO)	0.25	Cr	2	7.14
NH$_4$Cl	0.50	SiO$_2$	10	2.18
CH$_3$COOH	0.25	TiO$_2$	0.28	4.05
Bulk pH	3.0–3.5			
Operating Temperature	18–22°C			

samples were heated at 100°C, 150°C and 200°C for 1 hour in a furnace, before testing.

The coatings' resistance to red rust was investigated by exposing the samples to salt spray test according to BS 5466. The spray solution consisted of 5% NaCl (1.0M) and pH 6.0 at 38°C. The exposed surface of the sample was limited to 4cm². A periodic visual inspection was carried out in order to evaluate the initial onset and percentage coverage of red rust.

Electrochemical property measurements, first proposed by Stern and Geary[9], offer quick and reproducible methods to monitor corrosion behavior. After immersion in 1.0M, non-deaerated, NaCl solution, 30°C and pH 6.0, the corrosion current (I_{corr}) and corrosion potential (E_{corr}) were determined using EG&G Princeton Applied Research Model 352 SoftCorr ™ corrosion measurement and analysis software. This method was repeated in a solution of 0.5M NaCl and 0.5M NH$_4$Cl solution, 30°C and pH 8.0.

3 RESULTS AND DISCUSSION

3.1 Coating Characteristics

The zinc composites investigated in this study were prepared by electrodeposition from an electrolyte based on acetate. The composite coatings consisted of fine, lustrous crystals and showed such good coating adhesion that only small amounts of powder were detached in the 180° bending test. The optimal current density for achieving this was found in all cases to be in the range 2 to 4 A/dm².

X-ray diffraction measurements were carried out on a selection of the deposited composites. From the diffractograms obtained it was found, irrespective of plating conditions or particle type added, the resultant deposit was highly crystalline. A typical diffractogram is illustrated in Figure 1.

X-ray fluorescence spectrometry was used to determine the composition of the composites. In general the amount of particles incorporated within the deposit increased as the loading of the electrolyte was increased until a maximum was reached. Each particle type had an optimum plating current density that gave maximum inclusion. Figure 2 shows some SEM micrographs of the as-plated zinc composite with addition of different particle type[5]. In general the deposits contain randomly distributed hexagonal crystals with distinct boundaries. The particle size appears to cause an effect on the grain boundary size, the smaller the particle the finer the grain size becomes. Cross-sections of samples confirm this with the Zn-TiO$_2$ forming fine

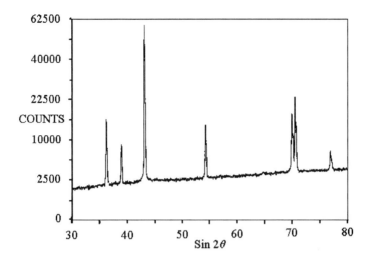

Figure 1 *X-Ray diffractogram of Zn-TiO$_2$ (2.6% wt) composite*

dendritic growth while, for same operating conditions, the Zn-SiO$_2$ has a much larger grain size. Introduction of particles into the electrolyte resulted in obvious morphological changes. In the case of addition of TiO$_2$ particles, the morphology of the coatings took the form of hexagonal plates (Figure 2d), characteristic of a pure zinc deposit (Figure 2a). On addition of Al$_2$O$_3$ particles, the hexagonal plate structure becomes more nodular (figure 2b). The

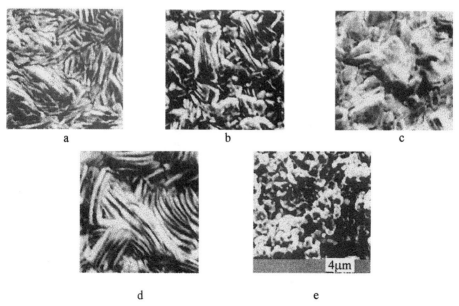

Figure 2 *SEM micrographs of as deposited (a) Zinc, (b) Zn-Al$_2$O$_3$, (c) Zn-SiO$_2$, (d) Zn TiO$_2$, (e) Zn-Cr. Operating current density of 2.5 A/dm^2*

micrograph shown in Figure 2c illustrates the effects of the addition of SiO_2 to the bath – irregular nodules larger than the ones obtained in the case of Al_2O_3. In the case of the addition of Cr particles (Figure 2e) much more compact, irregular nodules are formed.

3.2 Corrosion Resistance

The results from the accelerated corrosion tests are shown in Figures 3 and 4. The results show the time taken (hours) for 50% of the sample to be covered in red rust. All four MMC's provide an improvement to the rate at which red rust propagates by an average of 50%. In the case of $Zn-TiO_2$, the rate at which red rust coverage occurred was 90% better, under optimum conditions, when compared with the zinc. The order of ranking the corrosion performance of the four composites under investigation is:

$$Zn–Cr > Zn-TiO_2 > Zn–Al_2O_3 \text{ \& } Zn-SiO_2 > Zn$$

In the case of the Zn-Cr MMCs samples were found to maintain their integrity, with no samples exhibiting >50% red rust coverage having undergone the salt spray test for over 1000 hrs.

According to the salt spray test performance the optimum composition for each MMC's was found to be:

$$Zn-Cr \text{ (5–8 wt%)}, Zn-Al_2O_3 \text{ (4–6 wt%)}, Zn-SiO_2 \text{ (0.05–0.08 wt%)}, Zn-TiO_2 \text{ (2-3 wt%)}$$

The corrosion data extrapolated from the software generated Tafel plots are given in Tables 2, 3 and 4. The improved corrosion properties are given by the enhancement factor (EF). In

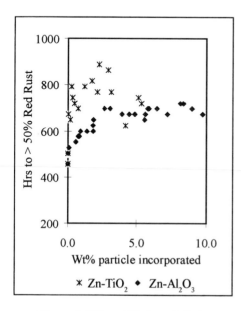

Figure 3 *Effect of TiO₂ and Al₂O₃ particle on corrosion behaviour*

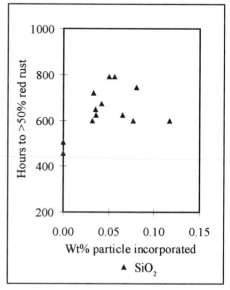

Figure 4 *Effect of SiO₂ particle on corrosion behaviour*

Table 2 *Corrosion data from potentiodynamic polarization (heat treatment in argon)*

Coating type	E_{corr} vs SCE mV, ±10	I_{corr} as-plated µA/cm²	EF	I_{corr} 100°C µA/cm²	EF	I_{corr} 150°C µA/cm²	EF
Zn	−1187	17.7	1.0	28.1	0.6	27.6	0.6
Zn-Al₂O₃ 5.0 wt%[†]	−1025	8.2	2.2	18.9	0.9	19.1	0.9
Zn-Cr 8.0 wt%[†]	−1136	2.3	7.7	12.3	1.4	14.6	1.2
Zn-SiO₂ 0.1wt%[†]	−995	11.7	1.5	23.5	0.8	20.1	0.9
Zn-TiO₂ 2.5 wt%[†]	−1033	3.6	4.9	16.0	1.1	15.4	1.1
Zn-Ni 13.8 wt%[‡]	−920	3	5.9	-	-	-	-
Zn-Co 1.0 wt%[‡]	−1060	7	2.5	-	-	-	-

[†] Electrolyte 1.0M NaCl (pH 6.0, 30°C), deposits 25µm thick.
[‡] Results quoted by Short et al [10]

the case of Zn-TiO₂ this would be:

$$\text{Enhancement factor (E.F.)} = \frac{I_{corr(Zn-as-plated)}}{I_{corr(Zn-TiO_2)}} = \frac{17.7}{3.6} \cong 4.9$$

The as-plated composites exhibited the following trend: (Zinc having greatest I_{corr} and thus poorest corrosion resistance)

$$Zn > Zn\text{-}SiO_2 > Zn\text{-}Al_2O_3 > Zn\text{-}Co \text{ (alloy)} > Zn\text{-}TiO_2 > Zn\text{-}Ni \text{ (alloy)} > Zn\text{-}Cr$$

Therefore the Zn-Cr composite offers superior corrosion resistance to that of Zn, Zn-Co, and slightly better than Zn-Ni. The composites have the added advantage that neither suffer from the problem of dezincification Zn-Ni alloys are prone to. This results in a much more noble

Table 3 *Corrosion data from potentiodynamic polarization (heat treatment in air)*

Coating type	E_{corr} vs SCE 100°C mV, ±10	I_{corr} 100°C µA/cm²	EF	E_{corr} vs SCE 150°C mV, ±10	I_{corr} 150°C µA/cm²	EF	E_{corr} vs SCE 200°C mV, ±10	I_{corr} 200°C µA/cm²	EF
Zn	−976	18.2	1.0	−917	20.9	0.8	−1008	23.1	0.8
Zn-Al₂O₃ 5.0 wt%[†]	−998	14.6	1.2	−955	18.5	1.0	−965	19.2	0.9
Zn-Cr 8.0 wt%[†]	−970	7.0	2.5	−926	12.8	1.4	−929	16.7	1.1
Zn-SiO₂ 0.1wt%[†]	−985	16.5	1.1	−967	20.4	0.9	−957	22.9	0.8
Zn-TiO₂ 2.5 wt%[†]	−905	7.6	2.3	−890	12.6	1.4	−977	17.4	1.0

[†] Electrolyte 1.0M NaCl (pH 6.0, 30°C), deposits 25µm thick.

Table 4 *Corrosion data from potentiodynamic polarization in alkali media*

Coating type	E_{corr} vs SCE mV, ± 10	I_{corr} as-plated $\mu A/cm^2$	EF
Zn	−1075	182.6	1.0
Zn-Al$_2$O$_3$ 5.0 wt%[*]	−1045	128.7	1.4
Zn-Cr 8.0 wt%[*]	−1015	111.6	1.6
Zn-SiO$_2$ 0.1wt%[*]	−1048	189.3	1.0
Zn-TiO$_2$ 2.5 wt%[*]	−1065	145.1	1.3

deposit that can in turn cause a reverse in the sacrificial nature of the material when the Ecorr approaches that of mild steel (E_{corr} −608mV). This would dramatically increase the corrosion of the steel the alloy was meant to protect.

In the case of heat treated samples it is beneficial to compare the composite coating with that of the original electrodeposited zinc. In the cases of MMC's that underwent heat treatment in the inert gas (argon), Table 2, all the four types exhibited a poorer corrosion resistance when compared to the original I_{corr} values. They still follow the same trend as before: (Zinc having greatest I_{corr} and thus poorest corrosion resistance)

$$Zn > Zn\text{-}SiO_2 > Zn\text{-}Al_2O_3 > Zn\text{-}TiO_2 > Zn\text{-}Cr$$

The MMCs that underwent heat treatment while exposed to air also show a decrease in the corrosion properties. If the value of EF is unity then the deposits' corrosion characteristics are equivalent to that of the as-deposited zinc, higher values indicate reduced corrosion rate. In the case of samples heated to 100°C all four have a better performance than the as-plated zinc under normal conditions. When the samples have been treated to 150°C the Zn-SiO$_2$ is the only composite of the group to give an EF of less than unity. At 200°C the Zn-Cr composite is the only remaining sample to give a value greater than unity. The Zn-TiO$_2$ composite provides, at 200°C a corrosion resistance on par with that of as-plated zinc.

Therefore the Zn-TiO$_2$ and Zn-Cr composites maintain a value greater than unity in terms of corrosion property enhancement. The other composites while deteriorating below the value given for the original electrodeposited zinc do show an improvement over zinc that has also been subjected to the same heat treatment. The cause for the reduction in the corrosion properties on heating has not been determined. One possible explanation would be that the particles within the matrix act as a focus for the production of micro-cracking within the matrix and thus the corrosion species will have easy access to the underlying metal. In the case of Zn-SiO$_2$ the value of I_{corr} shows an improvement. This may be due to stress relaxation within the deposit and heating at a higher temperature may improve this further. Further investigations are on-going to try to determine whether this is in fact the case.

The nobility of the composites is represented by the value of E_{corr}. Under acidic conditions, when exposed to air, the E_{corr} of the composites in general increases as the sample is heat treated. This suggests an oxidative reaction is taking place to form a more noble oxide layer on the surface of the sample. This may in some way reduce the effects of any thermal stresses introduced by the treatment.

Under alkaline conditions all of the composites perform poorly; although with the only exception of Zn-SiO$_2$ composite, still provide an improvement over the as-deposited zinc. This suggests composites of this nature are not suited for this type of environment. The corrosion properties of the composites follow the trend (Zn-SiO$_2$ having greatest I$_{corr}$ and thus poorest corrosion resistance):

$$Zn\text{-}SiO_2 > Zn > Zn\text{-}TiO_2 > Zn\text{-}Al_2O_3 > Zn\text{-}Cr$$

One noticeable difference from this trend when compared to the others is that the Zn-Al$_2$O$_3$ composite has improved its ranking compared to the others.

4 CONCLUSION

Zinc composites should offer the opportunity to provide good corrosion protection at temperatures in the range 150°C to 200°C when compared with a zinc only deposit. This development is seen as essential if zinc is to maintain its prime position as a sacrificial protective coating for steel within the automotive industry. As the present work suggests the order of preference for corrosion protection of the zinc composites investigated is:

$$Zn\text{-}Cr > Zn\text{-}TiO_2 > Zn\text{-}Al_2O_3 \text{ and } Zn\text{-}SiO_2 > Zn$$

In general the corrosion properties of zinc composites deteriorate after heat treatments. However, all composites performed much better than the zinc coating subjected to the same treatment.

The reasons for this deterioration in corrosion properties are under investigation. Zinc composites do not appear to be suitable for use in a marine alkaline environment. However, with the exception of the Zn-SiO$_2$ composite, all other composites still provide better corrosion performance over the as-plated zinc coating. The corrosion properties in this system are very poor and follow the order of preference for corrosion protection which is (greatest first):

$$Zn\text{-}Cr > Zn\text{-}Al_2O3 > Zn\text{-}TiO_2 > Zn\text{-}SiO_2 \text{ and } Zn$$

References

1. C. Y. Liue, J. W. Wang , Y. M. Peng, H. J. Chen, J. H. Shen and C. A. Hung , *MRL Bull. Rev. Dev.*, 1990, **4**, 31.
2. V. P. Khor'kov, I. G. Khabibullin and A. A. Yarullina, *Zashchita Metallov*, 1986, **22**, 101.
3. T. M. Duda, A. R. Kryuchkova, *Sverkhtverdye Materialy*, 1983, **5**, 38.
4. F. K. Sautter, *J. Electrochem Soc.*, 1963, **110**, 6.
5. Y. Suzuki, O. Asai, *J. Electrochem Soc.*, 1963, **134**.
6. K. Naitoh, S. Matsumura, K. Araki and T. Otaka, *Proc. Interfinish* , 1990
7. M. Kimoto, A. Yakawa and T. Tsuda, et al, *Metall*, 1990, **44**, 12.

8. Y. Shiohara and M. Abe, *NKK Corp. Technical Paper*.
9. M. Stern, *Corrosion* , 1958, **13**, 440.
10. N. R. Short, A. Abibsi, J. K. Dennis, *Trans. I.M.F.*, 1989, **67,** 73.

1.2.3
Aluminium-Magnesium Corrosion Resistant Ccoatings

K.R.Baldwin[1], R.I.Bates[2], R.D.Arnell[2] and C.J.E.Smith[1]

[1]STRUCTURAL MATERIALS CENTRE, DEFENCE RESEARCH AGENCY, FARNBOROUGH, HAMPSHIRE, UK

[2]RESEARCH INSTITUTE FOR DESIGN, MANUFACTURE AND MARKETING, UNIVERSITY OF SALFORD, SALFORD, UK

1 INTRODUCTION

Aluminium coatings are widely used for the corrosion protection of steel substrates and may be deposited using a variety of methods, including spraying, physical vapour deposition and electrodeposition from organic baths. The technology required to deposit aluminium coatings is mature and well established and is routinely employed on an industrial scale. In aerospace applications, PVD methods are generally preferred since they enable thin coatings to be deposited onto items requiring fine tolerances, e.g. on fasteners and other threaded parts. In recent years, there has been considerable interest in the use of sputtering techniques for the production of metal coatings for wear and decorative applications as well as for corrosion protection. One such technique is unbalanced magnetron sputtering (UMS), which has been successfully employed at the University of Salford to produce a wide range of pure metal, lamellar metal/metal oxide and graded alloy coatings[1,2]. In particular, UMS techniques have been employed to produce dense aluminium coatings for corrosion protection, that do not require post-plating peening treatments.

Recent work at DRA Farnborough[3] on as-plated UMS aluminium coatings has shown that they provide high levels of corrosion protection to steel substrates over a broad range of coating thicknesses. The UMS technique has been developed further at the University of Salford to produce aluminium-magnesium (Al-Mg) alloy coatings with controlled compositions and thicknesses[2]. The addition of moderate concentrations of Mg resulted in a marked improvement in the level of corrosion protection afforded by the as-plated Al coatings[3]. Despite these encouraging results, if Al-Mg alloy coatings are to find commercial use, it is important to demonstrate that they are receptive to conventional post-plating processes such as the application of chromate conversion treatments and that they can be heat-treated without a significant loss of corrosion resistance. In the present work, the corrosion behaviour of UMS Al-Mg coatings has been investigated in further detail with particular attention being paid to the effects of post-plating treatments.

2 PRODUCTION OF COATINGS

Pure Al and Al-Mg alloy coatings were deposited at the University of Salford using an unbalanced magnetron sputtering rig. Precise details of the unit and its operation have been described elsewhere [1,2]. Briefly, mild steel test panels of 1mm thickness (50mm x 50mm dimensions) were degreased and placed in a vacuum chamber. The chamber was then evacuated and backfilled with high purity argon gas. For pure Al coatings, the chamber contained a single pure Al target, whereas, for the alloys, the chamber was fitted with an additional target of pure Mg. Following sputter cleaning, a low-voltage was applied to the test panels to provide an external bias during deposition. The targets were then energised using an electrical current, the power levels being varied to obtain the required ratio of Al:Mg in the deposits. The coating process was continued for a sufficient time to allow the required thickness of deposit to form on the substrates. When the deposition process was completed, the coated panels were allowed to cool under vacuum. The composition of each deposit was determined using electron-microprobe analysis and the thickness of each coating was measured using a non-destructive eddy-current method. For the current work, single composition aluminium alloys containing up to 65 wt% magnesium were produced, with the thickness range being controlled between 2–25μm.

3 POST-DEPOSITION TREATMENTS

Following deposition, a range of single-stage or two-stage treatments was applied to separate batches of pure Al and Al-Mg coatings, as detailed in Table 1.

3.1 Single-Stage Treatments

3.1.1 Chromating. Chromate conversion coatings were applied to a batch of pure Al and Al-Mg coatings using a proprietary chemical treatment, to Def. Stan 03-18[4]. Alloy coatings containing 15 wt% Mg or less, were readily chromated, producing films with a typical yellow/orange finish. However, for alloy coatings containing over approximately 15 wt% Mg, the chromate treatment either completely failed to generate films or produced unsatisfactory, loosely adherent, brown deposits. The application of chromate treatments developed specifically for magnesium-rich alloys[5] was also unsuccessful with alloy coatings containing over 15 wt-% Mg. In the present work, therefore, corrosion studies of chromated coatings were restricted to Al alloys containing approximately 15 wt% Mg or below.

3.1.2 Heat-Treatment. A separate batch of specimens were heat-treated in laboratory air at 220°C for 8 hrs. The heat-treatment, applied to batch 3 specimens, did not change the visual appearance of the coatings.

To examine the effect of the heat-treatment on the structure of Al-Mg coatings, X-ray

Table 1 *Post-deposition treatments applied to UMS aluminium coatings*

Batch Number	Type of process	Stage 1	Stage 2
1	-	None	None
2	Single-stage	Chromated	None
3	Single-stage	Heat treated	None
4	Two-stage	Heat treated	Chromated
5	Two-stage	Chromated	Heat treated

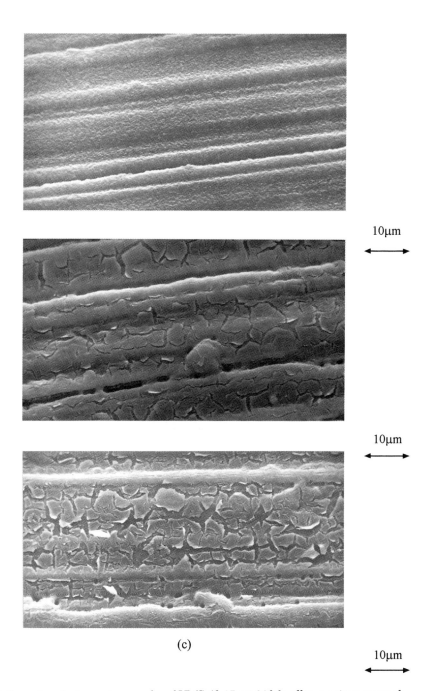

10μm

10μm

(c)

10μm

Figure 1 *Scanning electron micrographs of UMS Al-17 wt.% Mg alloy coatings on steel; (a) as-plated,(b) plated and chromated and (c) plated, chromated and baked, each taken at 90° to the surface. As-plated coating thickness: 12μm*

diffraction (XRD) analysis was carried out on as-plated Al alloys. Two Al alloy coatings, containing 2.9 and 16.0 wt.% Mg, were analysed before and after heat-treatment at 220°C for 8hrs. A detailed analysis of the XRD spectra is outside the scope of this paper, although the phases present in the alloys have been described elsewhere[6,7]. In the present work, analysis of the XRD spectra prior to baking indicated that the structure was single phase for both compositions. No evidence of the presence of the intermetallic β-phase was found in the alloy containing 16 wt.% Mg even though the equilibrium phase diagram[7] indicates that this phase should be present at this concentration. Detailed comparisons of the XRD patterns of the alloys were made with those obtained after heat-treatment and were found to be identical. This indicated that the structures of the alloys were not significantly affected by the application of sustained elevated temperatures.

3.2 Two-Stage Treatments

The two-stage treatments detailed in Table 1 were applied to two separate batches of pure Al and Al-Mg alloy coatings. One batch of alloy coatings, containing 15 wt.% Mg or less, was heat-treated at 220°C for 8hrs prior to the application of the chromate chemical treatment. The coatings were allowed to cool in air, to 20°C, before immersion in the chromating bath. The chromating characteristics of the coatings did not appear to be affected by the initial heat treatment. For batch 5 coatings, the as-plated coatings were chromated and then heat-treated at 220°C for 8hrs. The heat treatment of the previously chromated specimens caused a change in the appearance of the surface finish from pale yellow/orange to an orange/purple colour.

4 APPEARANCE AND STRUCTURE

The range of coatings produced by magnetron sputtering (0–65 wt.% Mg) was generally bright in appearance and exhibited excellent adhesion to the substrate. Scanning electron microscopy was used to examine the structures of selected Al-Mg coatings.

Figure 1 shows the surface morphologies of an Al–17 wt-% Mg alloy; as-plated, chromated, and chromated and heat-treated. The SEM of the as-plated coating shows the typical dense, non-columnar, morphology of UMS coatings[2,3]. The application of the chromate chemical treatment formed a layer on the surface of the metal coating which covered many of the minor peaks and troughs observed on the as-plated specimen. The subsequent baking of the plated and chromated specimen caused the chromated layer to dry and crack, exposing much of the underlying metal coating.

Table 2 *Corrosion current density values determined for UMS coatings in 600 mmol λ^{-1} NaCl solution at 25°C*

Coating	Compositions (wt.% Mg)	Corrosion current density ($\mu A\ cm^{-2}$)		
		As-plated	Chromated	Heat-treated
Pure Al	-	0.14	0.04	0.15
A-5 wt.% Mg	5.4 ± 1.2	0.18	0.02	0.05
Al-10 wt.% Mg	10.0 ± 1.2	1.26	0.08	0.04
Al-15 wt.% Mg	15.3 ± 2.6	1.92	0.09	0.05

Table 3 *Open-circuit potential values determined for UMS coatings in 600mmol λ^{-1} NaCl solution at 25°C*

Coating	Compositions (wt.% Mg)	Open-circuit potential (V vs SCE)		
		As-plated	Chromated	Heat-treated
Pure Al	-	− 0.82	− 0.78	− 0.79
Al-5 wt.% Mg	5.4 ± 1.2	− 0.89	− 0.79	− 0.90
Al-10 wt.% Mg	10.0 ± 1.2	− 0.93	− 0.82	− 0.92
Al-15 wt.% Mg	15.3 ± 2.6	− 0.97	− 0.85	− 0.95

5 CORROSION STUDIES

The corrosion resistance of a metal coating is a function of two main properties of the coating; firstly, its ability to act as a barrier layer by isolating the underlying steel substrate from the aggressive environment, and secondly, its ability to afford sacrificial protection once the coating

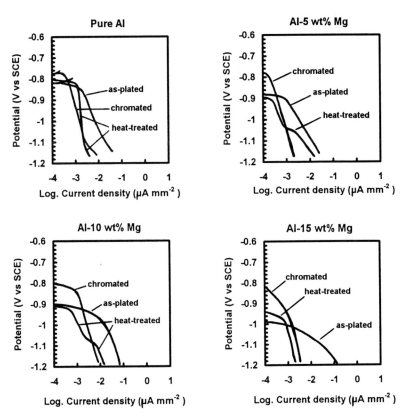

Figure 2 *Cathodic potentiodynamic polarization sweeps of UMS pure Al and Al-Mg alloy coatings in aerated 600mmol λ^{-1} NaCl at 25°C. Sweep rate 0.125mV s^{-1}*

Table 4 *Predicted and Actual time to red-rust values for UMS coatings.*

Coating	T_{RR} contribution (hrs)			T_{RR} predicted (hrs)	T_{RR} actual (hrs)
	As-plated	Heat-Treatment	Chromating		
Pure Al	390	16	234	640	622
Al–5% Mg	435	167	1149	1751	1452
Al–15% Mg	707	345	644	1696	1303

has been breached. The coating will cease to afford full protection to the substrate following the loss of both barrier and sacrificial properties. At that point, the steel will then be free to corrode, forming characteristic iron-oxide (Fe_2O_3) corrosion products (red-rust).

In the following section, the barrier and sacrificial properties of Al-Mg coatings have been examined in isolation, by electrochemical techniques. The overall corrosion behaviour of the coatings on steel has then been evaluated for a range of coating thicknesses, using neutral salt fog tests. For the electrochemical tests, thick coatings ($20\mu m$), were employed to minimise the possibility of interactions with the substrate.

5.1 Electrochemical Investigations

5.1.1 Barrier Properties. A previous study at DRA Farnborough[3], has shown that the barrier properties of metal coatings may be conveniently examined using cathodic potentiodynamic polarisation techniques. In the present work, cathodic sweeps have been carried out on pure Al and alloy coatings containing up to 15 wt.% Mg, for single stage treatments only (batches 1 to 3).

Figure 2 shows the cathodic sweeps obtained after 30 minutes in aerated 600mmol λ^{-1} NaCl solution. In each case, the general form of the polarisation curves was similar in that, at moderate overpotentials, only relatively small increases in current density were obtained for large increases in the polarization overpotential. This type of polarisation behaviour is usually associated with cathodic reduction processes that are predominantly under diffusion control e.g. oxygen reduction. The corrosion current density, i_{corr}, of each coating was determined from the cathodic polarization curves, using a Tafel extrapolation technique[8]. The values obtained are given in Table 2.

The data obtained for the as-plated specimens show that, as the Mg content was raised, the corrosion current density increased. This is consistent with earlier work[3] and suggests that as the Mg content was increased, so the barrier properties of the alloy coatings declined. Table 2 shows that both post-plating treatments had a significant beneficial effect on the corrosion resistance of the coatings. The chromate treatment significantly reduced the corrosion current density of each coating, indicating that the chromate conversion layer acted to improve the barrier properties of each coating evaluated. The heat-treatment also appeared to significantly

Table 5 *Effect of heat treatment on selected chromated coatings*

Coating	Time to red-rust (hrs)	
	Chromated only	Chromated then heat-treated
Pure Al	624	456
Al-5 % Mg	1584	719
Al-15 % Mg	1560	822

improve the barrier properties of the alloy coatings but had little effect on pure Al.

5.1.2 Sacrificial Properties. Earlier work at DRA Farnborough[3,9] has shown that information relating to the sacrificial properties of metal coatings may be obtained from open-circuit potential measurements. In this study, the open-circuit potentials of selected alloy coatings were measured after 30 minutes in quiescent 600mmol λ^{-1}NaCl solution. The values obtained are shown in Table 3.

For as-plated Al coatings, the open-circuit potential was found to decrease as the concentration of the more active Mg metal was increased, which is in accordance with previous work[3]. Table 3 shows that the heat-treatment had no significant effect on potential. In contrast, although the chromating of as-plated specimens caused a negative shift in potential as the Mg level was raised, the effect was considerably less pronounced than with the as-plated coatings.

5.2 Neutral Salt Fog Tests

The electrochemical techniques have been used to examine the barrier and sacrificial properties of the UMS coatings in isolation. In this part of the study, the overall corrosion

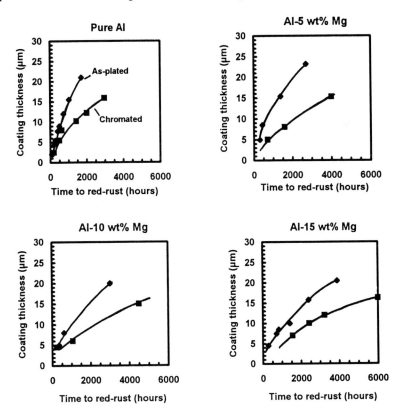

Figure 3 *Relationship between coating thickness and time to first red-rust for UMS Al and Al-Mg alloy coatings on steel, on exposure to continuous neutral salt fog. Symbols: diamonds-as-plated specimens, squares-plated and chromated specimens*

behaviour of the coatings on steel was determined using neutral salt fog tests. The performance of the coatings in these tests was assessed by determining the time taken for red-rust, T_{RR}, to appear on the coated panel surfaces. T_{RR} will be related to the time taken for the loss of both barrier and sacrificial corrosion protection.

5.2.1 Single-Stage Treated Coatings. Neutral salt fog tests were conducted for coating batches 1 to 3, in accordance with ASTM B117[10]. For batch 1, coatings containing 0, 5, 10 and 15 wt.% Mg were evaluated and the relationship between T_{RR} and coating thickness is shown by Figure 3. The corrosion resistance of the as-plated coatings was found to increase in a near-linear fashion as the thickness was increased. The main type of coating corrosion observed in the salt fog environment was pitting attack, with some staining and generation of white corrosion products. Figure 3 indicates that the chromate treatments (batch 2) significantly improved the corrosion performance of the coatings evaluated.

Figure 4 shows the effect of alloy composition on corrosion performance for as-plated and chromated coatings, for a single coating thickness of 8μm. The T_{RR} values for as-plated coatings were found to increase as the Mg content was raised, until a peak in performance was obtained for alloys containing approximately 20 wt.% Mg. Further increases in Mg content were of no further benefit but instead caused a decline in corrosion resistance. Figure 4 also shows the effect of the chromating on T_{RR}, for coatings containing 15 wt.% Mg or less. The chromate treatment significantly improved corrosion resistance, and the effect became generally more pronounced as the alloy Mg content was raised.

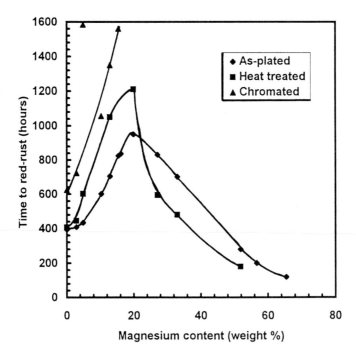

Figure 4 *Relationship between time to first red-rust and composition for 8μm UMS Al-Mg alloy coatings on steel, on exposure to continuous neutral salt fog*

The effect of heat treatment (batch 3) on T_{RR} was determined for a single coating thickness of 8μm and the data obtained are also shown in Fig. 4. For Al alloy coatings containing approximately 20 wt-% Mg or below, the bake at 220°C for 8 hrs caused a significant improvement in corrosion resistance although the compositional peak in performance remained unchanged at 20 wt-% Mg. In contrast, for the high-Mg alloys, above approximately 20 wt.% Mg, the heat treatment caused a decline in performance.

5.2.2 Two-Stage Treated Coatings. The effect of heat-treatment prior to chromating (batch 4) was established for pure Al and two of the alloys (5 and 15 wt.% Mg), again for a single coating thickness of 8μm, and the data obtained are given in Table 4.

The heat treatment prior to chromating was found to have a slightly detrimental effect on corrosion resistance, although the actual T_{RR} values obtained represent a significant improvement over the as-plated coatings. The data obtained from the single-stage treatments (section 5.2.1) were used to predict T_{RR} values for the coatings that had been heat-treated and then chromated.

The contribution made by chromating was determined by subtracting the T_{RR} values from batch 2 from those obtained in batch 1. For example, for pure Al, the T_{RR} for as-plated coatings was 390 hrs, whereas the chromated coatings gave a value of 624 hrs, indicating that chromating increased T_{RR} by 234 hrs. Similar calculations were made for heat-treated coatings and Table 4 shows the T_{RR} values, both predicted and actual for the batch 4 coatings. Table 4 shows that the actual T_{RR} values obtained from the neutral salt fog tests were slightly lower than those predicted by simple addition of the separate T_{RR} contributions. This suggests that the heat-treatment may have reduced the effectiveness of the chromate conversion treatment, although the effect was relatively small.

The batch 5 specimens consisted of as-plated coatings of a single thickness (8μm) which had been chromated and then heat-treated. The T_{RR} values obtained for the batch 5 samples are shown in Table 5. T_{RR} data for chromated only coatings (batch 2) are included for comparison.

The data in Table 5 show that the heat-treatment of chromated coatings (which had not been previously heat-treated) had a detrimental effect on corrosion performance in the neutral salt fog environment. This effect was more pronounced for the Al-Mg alloys (5 wt.% and 15 wt-% Mg) than for pure Al.

6 DISCUSSION

6.1 Barrier Properties of Coatings

6.1.1 General. The main factors that determine the ability of a coating to act as a barrier layer are coating corrosion rate, coating microstructure and mode of coating corrosion. The microstructure is an inherent property of the coating layer, determined largely by the mode of deposition, whilst the corrosion rate and mode of corrosion are partly dependent on the nature of the corrosive environment itself.

The coating microstructure is important since it will determine how effectively a metal layer can isolate the substrate from the corrosive environment. Two extreme cases are the coating with an open, porous structure, often found with electrodeposits, and the dense morphologies with no significant porosity, such as the magnetron sputtered coatings evaluated in the present work (Figure 1). The magnetron sputtered coatings will clearly be effective as barrier layers since the population of pre-existing pathways leading from the outer coating surface down to the substrate surface will be very much lower than those with a more open, porous, structure.

For a coating that is essentially non-porous, and corrodes in an entirely uniform manner, the time taken for coating loss, and hence time to red-rust, could be calculated simply from corrosion rate data. However, in practice, this situation is rarely observed since the corrosion of coating layers, particularly those with significant porosity, is nearly always non-uniform. Thus the mode of corrosion is also an important factor in determining the life of a barrier layer. In the case of aluminium coatings, localised corrosion is often the most significant mode of attack[3]. In the present work, pitting corrosion was observed in the neutral salt fog environment and hence exposure of the underlying steel is likely to occur at an earlier stage than if corrosion were more uniform in nature. The effects of both microstructure and the mode of corrosion on the barrier properties of coatings are difficult to quantify. However, in the present work, semi-quantitative data on the barrier properties of the coatings were obtained from the cathodic potentiodynamic polarization sweeps, which are discussed further below.

6.1.2 As-Plated Coatings. Aluminium-base coatings have an advantage over most of the other main sacrificial metal coatings in that the air-formed aluminium oxide (Al_2O_3) film provides a distinct layer with inherent corrosion resistance. For aluminium-base coatings, the oxide layer will provide the first barrier against attack. Clearly the inclusion of alloying additions in the underlying metal layer which act to disrupt the growth of the aluminium oxide film will reduce its efficiency to act as a barrier. The cathodic sweeps conducted on as-plated coatings showed that the corrosion current density of pure Al was shifted to higher currents as the Mg level was raised (Table 2). It is probable that alloying additions of Mg to UMS Al coatings disrupted the surface oxide film and weakened its protective effect, as described in more detail in an earlier publication[3].

6.1.3 Chromated Coatings. The cathodic sweeps showed that the application of a chromate conversion layer to pure Al and Al-Mg alloy coatings significantly improved the barrier properties (Table 2). Figure 1(b) showed the presence of the chromate conversion layer on the surface of an Al-Mg coating. Chromate conversion coatings are thought to consist mainly of mixed chromium/metal oxide films, typically 0.1-1mm in thickness, which contain significant concentrations of hexavalent chromate (CrO_4^{2-}), a known corrosion inhibiting species[11-15]. In the present work, it is envisaged that the chromate conversion coatings improved barrier properties in two main ways; firstly, by helping to isolate the underlying metal coating from the environment by functioning as an additional barrier layer, and secondly, in the presence of moisture, soluble chromate species are thought to leach out from the conversion layer and inhibit corrosion in areas where the conversion layer is incomplete or damaged. The conversion layer will be effective until the soluble chromate in the film becomes exhausted.

6.1.4 Heat-Treated Coatings. The cathodic current density data presented in Table 2 showed that the heat-treatment of UMS Al alloy coatings containing 15 wt.% Mg or less resulted in an improvement in barrier properties, to a similar extent as that found with chromating. It is proposed that the heat treatment caused some sealing of pre-existing porosity in the surface aluminium oxide film on the metal coating surface, possibly accompanied by oxide thickening[17]. The reduction in porosity would have promoted the formation of a more efficient barrier layer.

Heat-treatments were also applied to UMS coatings which had previously been chromated. Figure 1(c) showed that the chromated layer on specimens which had been subsequently heat-treated, had become severely cracked. This indicated that the heat-treatment of chromated specimens reduced the barrier properties of the chemical conversion layer, although soluble chromate species are still likely to be present. Brumer[16] identified similar effects, although it was found that chromate conversion coatings could be heated to approximately 80°C without a significant loss of performance.

6.2 Sacrificial Properties of Coatings

6.2.1 General. Despite the good barrier protection afforded by many metal coatings, continued exposure to a corrosive environment will eventually lead to a loss of barrier protection and the underlying steel substrate will become exposed to the corrodent. At that point, further corrosion protection will then depend on the ability of the remaining metallic coating layer to sacrificially protect the areas of exposed steel.

A previous study at DRA Farnborough[18], has shown that the open-circuit potential of a metal coating must be at least approximately 50mV negative to bare mild steel, to drive the steel below its protection potential and afford adequate cathodic protection to the steel. If the steel is above the protection potential, it will be free to corrode and form red-rust corrosion products. The open-circuit potential of bare mild steel in quiescent 600mmol λ^{-1} NaCl solution, is -710mV (SCE)[19]. Since the as-plated UMS pure Al coatings evaluated in the present work were approximately 110mV negative to this value (Table 3), then they should provide full protection to the steel substrate at any breach in the coating. In practice, however, it is often found that the sacrificial protection afforded by pure Al coatings is inadequate due, in part, to their tendency to passivate[3].

The potential adopted by a binary alloy will be at a point between the potentials of the parent metals[20]. The open-circuit potential of pure Mg in 600mmol λ^{-1} NaCl solution is approximately -1.9 V (SCE)[3]. Accordingly, the addition of Mg to the sputtered Al deposits produced alloys that adopted open-circuit potential values that were negative to that of pure Al but were positive with respect to that of pure Mg. The addition of Mg to Al will improve the sacrificial properties of the alloy coatings and, provided the electrolyte conductivity is sufficient, will allow an increasingly larger area of steel to be exposed before red-rust formation can occur. However, a point will be reached, when the Mg concentration is at a certain level, when galvanic corrosion of the coating will become so severe that any improvement in sacrificial properties will be out-weighed by the rapid rate of coating loss.

6.2.2 Effects of Post-Plating Treatments. The open-circuit potential data (Table 3) suggested that the post-plating treatments may also influence sacrificial properties as well as barrier properties. The chromating of pure Al and Al-Mg alloy coatings shifted the potentials of the coatings in a noble direction, showing that the chromate conversion treatment would have had a detrimental effect on sacrificial properties. However, any loss of sacrificial properties is likely to be out-weighed by the presence of soluble chromate species from the chemical conversion layer, since, if the underlying steel was exposed, chromate ions would leach into the damaged region and inhibit corrosion. Table 3 shows that the heat treatment of pure Al and Al-Mg coatings would have had little effect on sacrificial properties.

6.3 Corrosion Behaviour of Coated Steel

In the present work, the time taken for red-rust to appear on the panel surfaces, T_{RR}, was used as a means of assessing overall corrosion performance. As indicated above, T_{RR} will be a function of both the barrier and sacrificial properties of the coatings.

In the neutral salt fog tests, it was found that the level of corrosion protection afforded to steel by the as-plated UMS Al coatings increased sharply as the Mg content was increased until a peak in performance was observed for alloys containing approximately 20 wt.% Mg (Figure 4). It is proposed that the increase in corrosion resistance observed in the neutral salt fog tests, as the Mg content was raised, was mainly due to a gradual improvement in sacrificial properties. Although the addition of Mg reduced the barrier properties of the UMS Al coatings, it has been established[3] that the effect does not become significant until the Mg content is above approximately 20 wt.%. The decline in the level of corrosion protection afforded by the as-plated Al coatings observed when the alloy Mg level exceeded 20 wt.% can be attributed to excessive galvanic corrosion as the coatings became more active, and the rapid loss of barrier properties.

The application of chromate conversion coatings to pure Al and Al–Mg alloys resulted in a significant improvement in the ability of the metal coatings to protect steel. The effect of the chromate conversion coatings was a complex one but clearly improved barrier properties and the presence of soluble chromate species were significant factors in their ability to protect both the aluminium layer and the underlying steel.

The neutral salt fog tests showed that the heat-treatment of as-plated pure Al and Al-Mg coatings also resulted in a significant improvement in the corrosion resistance of the coatings, provided the Mg content was maintained below approximately 20 wt-%. The electrochemical data showed that the improvement in corrosion resistance caused by heat treatment was due to an improvement in the barrier properties, or more specifically, the protective properties of the aluminium oxide surface film. Once the aluminium oxide film had been breached, the heat-treated coatings would have behaved very much like the as-plated coatings, which accounts for the peak in corrosion resistance remaining largely unchanged at 20 wt.% Mg. Fig. 4 showed that the heat-treatment of Al alloys containing over approximately 20 wt-% Mg caused a steeper decline in performance than that observed for the as-plated coatings. It is not clear why the heat-treatment should have had this effect since the XRD analysis ruled out changes in structure as a cause of any decline in corrosion performance. One possibility is that the formation of significant concentrations of essentially non-protective magnesium oxides in the surface coating layer caused a further loss of barrier properties.

7 CONCLUSIONS

1. Neutral salt fog tests conducted on as-plated UMS Al alloy coated steel sheet showed that the corrosion resistance increased as the Mg content was raised, until an optimum level of performance was attained for coatings containing approximately 20 wt.% Mg. Further increases in Mg content caused a decline in corrosion resistance.
2. Al alloy coatings containing 15 wt.% Mg or below were successfully chromated using a proprietary chemical treatment, although those coatings containing over 15 wt.% Mg could not be chromated. Neutral salt fog tests showed that the chromate conversion treatments applied to as-plated Al alloy coatings containing 15 wt-% Mg or below, provided a significant improvement in corrosion resistance.

3. Neutral salt fog tests showed that the heat treatment of as-plated UMS Al and Al alloys containing 20 wt.% Mg and below, caused a significant improvement in corrosion resistance compared with coatings which had not been heat-treated. The heat-treatment of Al alloys containing over 20 wt.% Mg had a detrimental effect on corrosion resistance.

Acknowledgement

The authors acknowledge the assistance of Mr R. Luke with some of the experimental work.

References

1. D. P. Monaghan, D. G. Teer, P.A.Logan, K. C. Laing, R.I.Bates and R.D.Arnell, *Surface and Coatings Technology*, 1993, **60**, 592.
2. R. I. Bates and R. D. Arnell, *Surface and Coatings Technology*, 1994, **68/69**, 686.
3. K.R.Baldwin, R.I.Bates, R.D.Arnell and C.J.E.Smith, *Corrosion Science*, 1996, **38**,155.
4. 'Chromate conversion coatings (chromate filming treatments) for aluminium and aluminium alloys', Def. Stan 03-18, MoD, Directorate of Standardization, Kentigern House, Brown Street, Glasgow, UK.
5. 'Protection of Magnesium-Rich Alloys Against Corrosion', DTD 911C, MoD, Specifications Branch, Structural Materials Centre, DRA Farnborough, Hants, UK, 1963.
6. 'Binary alloy phase diagrams', ed. T.B.Massalski, Vol.1, ASM International, 1990, Vol. 1, p. 169.
7. R. D. Arnell and R. I. Bates, *Vacuum*, 1992, **43** (1–2), 105.
8. F.Mansfeld in, 'Electrochemical Techniques for Corrosion' (ed. R.Baboian), NACE, Houston, Texas, 1977, p. 18–26.
9. K. R. Baldwin, C.J.E.Smith and M.J.Robinson, *British Corrosion J.*, 1994, **29**, 299.
10. 'Standard Method of Salt Spray (Fog) Testing', ASTM B117, American Society for Testing and Materials, Race Street, Philadelphia, 1981.
11. H. A. Katzman, G.M.Malouf and G.W.Stupian, *Appl. Surface Science*, 1979, **2**, 416.
12. B. R. W.Hinton, *Metal Finishing*, September 1991, 55.
13. Z. Yu, H. Ni, G. Zhang, Y. Wang, S. Dong and G. Zhao, *Appl. Surface Science*, 1992, **62**, 172.
14. F. W. Lytle, R. B.Greegor, G. L.Bibbins, K.Y. Blohowiak, R. E. Smith and G. D. Tuss, *Corrosion Science*, 1995, **37**, 349.
15. S. M. Cohen, *Corrosion (NACE)*, 1995, **51**, 71.
16. H. Brumer, 'Product Finishing Directory', 1993, p. 206.
17. Y. Mukai, K. Hatanaka and M. Fukui, 'Study of oxide film properties of Al-Mg alloys', in: Proc Conf. Sur/Fin '91, AESF, Toronto, Canada, 24–27 June 1991, p. 763.
18. K. R.Baldwin, M. J.Robinson and C. J. E.Smith, *Corrosion*, 1995, **51**, 932.
19. K. R.Baldwin, M. J.Robinson and C. J. E. Smith, *British Corrosion J.*, 1994, **29**, 299.
20. H. Kaiser, 'Corrosion Mechanisms' (ed. F.Mansfeld), Marcel Dekker Inc., 1987, p. 85.

1.2.4

Effect of Plasma and Conventional Gas Nitriding on Anodic Behaviour of Iron and Low-Alloy Steel

J. Mankowski and J. Flis

INSTITUTE OF PHYSICAL CHEMISTRY OF THE POLISH ACADEMY OF SCIENCES
01–224 WARSAW, POLAND

1 INTRODUCTION

Nitriding of ferritic steels is performed either in conventional processes in the atmosphere of ammonia or in a glow discharge atmosphere of nitrogen/hydrogen mixtures, usually in the temperature range of 550–580°C[1-4]. The primary objective of this thermochemical treatment is to improve mechanical properties by increasing fatigue and yield strength, hardness and wear resistance. Mechanical improvement is usually accompanied by an increase in corrosion resistance[5].

Nitriding produces a compound zone at the outermost surface and a diffusion zone below it. On iron the compound zone can be composed of γ'-phase (Fe_4N), ε-phase ($Fe_{2-3}N$) or a two-phase mixture of $\gamma' + \varepsilon$. Diffusion zones are much thicker than the compound ones; in diffusion zones nitrogen exists in the form of fine dispersions of nitride precipitates and also in a solid solution in the metal matrix.

Hardness resulting from iron nitrides is too low for engineering applications, therefore, steels used for nitriding contain effective nitride-forming elements such as Al, Ti, Cr, Mo, and V.

The diffusion zones determine hardness and fatigue strength of the nitrided layers, whereas the compound zones are decisive for tribological and corrosion characteristics[6].

Typically, compound zones have significantly higher corrosion resistance than that of the matrix, whereas diffusion zones have lower resistance than matrix. The ε nitrides usually show higher resistance than that of γ' nitrides owing to a different structure and/or higher nitrogen content. In 0.1 M H_2SO_4 the γ' and ε nitrides on iron corrode more slowly than iron, while the diffusion zone corrodes much more quickly[7]. In 0.01 M H_2SO_4 + 0.99 M Na_2SO_4 (pH 2.8) an homogeneous alloy of Fe–0.9 wt.% N showed a significantly higher tendency for passivation than iron[8]. Higher corrosion resistance of ε nitride in comparison to an untreated matrix was also shown for carbon steel in neutral solutions of 0.1 M NaCl and 0.05 M K_2SO_4 [9] and for iron in 0.05 M NaCl in which the tendency for passivation increased in the order $Fe < Fe_4N < Fe_{2.3}N$[10]. In 0.9 M NaCl solution the ε- and γ'-nitride layers decrease corrosion rate by about one order of magnitude, and ε-carbonitride layers decrease it by two orders of magnitude in comparison to the matrix material[11]. In low-alloy steels with 2% Cr and 5% Cr the best ability to passivate in the NaCl solution was shown by the ε-carbonitride phase with the concentration ratio C/(N+C) of 0.02 to 0.2[12].

Superior corrosion resistance of ε nitride in comparison to untreated matrix and to γ nitride was also observed for Fe-Cr alloys. Vorhoshkov and Sedloyev [13] found that for steels with up to 5 wt.% Cr corrosion resistance in neutral solutions of 0.1 M NaCl or 0.05 M Na_2SO_4 increased in the order: untreated steel $< \gamma' < \epsilon < \epsilon + \gamma'$. Similarly, Ibendorf and Schröter [14] reported that for iron and Fe-Cr alloys with up to 13 wt.% Cr corrosion resistance in an acidic 0.025 M H_2SO_4 + 0.05 M Na_2SO_4 solution increased in the order: untreated matrix $< \gamma' < \epsilon$. The high corrosion resistance of ϵ nitride was ascribed to its high passivation tendency. [14]

In the present work corrosion behaviour of iron and Cr3Mo steel was examined after plasma nitriding and conventional gas nitriding in ammonia. The composition of plasma nitrided layers was varied by using different nitrogen contents in the nitriding gas mixture and different duration of treatment. It was attempted to relate corrosion behaviour at various depths of nitrided layers to their chemical composition.

2 EXPERIMENTAL PROCEDURE

Two materials were used in this investigation:

i.   Armco iron (C 0.016, Mn 0.13, Si 0.001, S 0.009, and P 0.006 wt.%); it was normalised from 940°C;
ii. sorbitic Cr3Mo steel (C 0.41, Cr 3.00, Ni 0.09, Mn 0.68, Si 0.33, P 0.026 and Mo 0.64 wt.%); it was oil quenched from 920°C and then tempered for 2 h at 620°C.

Samples were in form of 3mm thick discs of 20mm diameter. They were polished to a final 600 grit with SiC emery paper prior to treatment.

Nitriding was performed at 540°C in an atmosphere of ammonia for 9 h, and in the glow discharge regime in the N_2/H_2 gas mixture with 25 or 80 vol.% N_2 for 6 or 9 h under a pressure of 670 Pa. In the Cr3Mo steel the outer zone obtained in the low N_2 atmosphere consisted of the γ' nitride (M_4N) and was 4 mm thick after 9 h, whereas in the high N_2 atmosphere it consisted of the ϵ ($M_{2-3}N$) + γ' nitrides and was 17μm thick after 6h [15]. Diffussion zone contained precipitates of Fe_3C and CrN; it extended to the depths of 160μm and 180μm for the low and high N_2 atmospheres, respectively. In iron, diffusion zone contained precipitates of Fe_4N and $Fe_{16}N_2$.

Microhardness was measured on cross-sections of the samples.

Depth profiles of the nitrogen and carbon concentration were obtained by glow discharge optical emission spectroscopy (GDS).

Corrosion behaviour was examined at 25°C in non-deaerated 0.05 M Na_2SO_4 solution, acidified with H_2SO_4 to pH 3.0. Polarization curves were measured at a potential scan rate of 1 mV s⁻¹, starting from a potential ~ 50 mV more negative than the open circuit value in the active region. Potentials were measured relative to a sulphate electrode and reported with respect to a saturated calomel electrode (SCE). Measurements were made on as-nitrided surfaces and for various depths obtained by abrasion.

3 RESULTS AND DISCUSSION

3.1 Microhardness and Elemental Concentration

Microhardness depth profiles (Figure 1) indicate that in Cr3Mo steel the gas nitriding

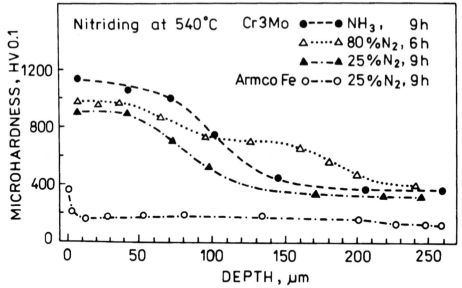

Figure 1 *Microhardness vs. depth for Cr3Mo steel after nitriding in ammonia and in plasma of N_2/H_2 gas with 25% N_2 and 80% N_2, and for iron after plasma nitriding*

produced layers of higher hardness than that obtained by plasma nitriding. The use of the high-nitrogen gas mixture for plasma nitriding resulted in an increase of the layer thickness, but hardness in the deeper layer was lower than in the outer one.

Nitriding of iron almost did not affect its hardness. An increase in the hardness occurred in a thin superficial layer, whereas for the diffusion layer it was only slightly above the hardness for the substrate. This is reasonable in view of the lack of strong nitride-forming elements in Armco iron.

Depth profiles of concentration of nitrogen and carbon for iron and Cr3Mo are shown in Figure 2 and Figure 3, respectively. On the basis of the weight concentrations obtained from GDS, atomic fractions of nitrogen were calculated for the outer zones and corresponding compositions of the nitrides in these zones were evaluated. These compositions are presented here as M_xN, where "x" is the evaluated factor.

From the nitrogen concentration profiles for iron (Figure 2) it follows that after nitriding in the 80% N_2 gas the outermost zone was composed of $Fe_{2.3}N$ which can be assigned to ε nitride, while the deeper one was composed of $Fe_{3.3}N$ which can be assigned to γ' nitride. After nitriding in the 25% N_2 gas there formed only one zone of an average composition $Fe_{3.3}N$.

At the end of the γ' zone there occurred a strong accumulation of carbon, being for the 80% N_2 gas twice as high as for the 25% N_2 gas. Evidently, it was due to the displacement of carbon by nitrogen, but it is not clear why at the depths closer to the surface the concentration of carbon was higher than that in the substrate. Possibly, carbon might have entered the samples from the nitriding reactor. Earlier an accumulation of carbon was reported for nitrocarburised iron-carbon alloys and low alloy steels[4].

Accumulation of carbon at the end of the nitride zone also occurred in Cr3Mo steel (Figure 3), however, it was less pronounced than that in iron. Similarly as in iron, the carbon

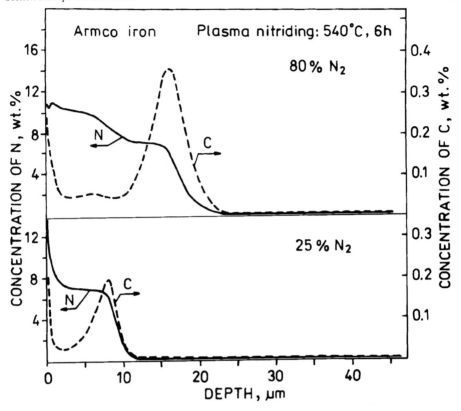

Figure 2 *GDS depth profiles for nitrogen and carbon in plasma nitrided Armco iron*

concentration near the surface was higher than in the substrate. Concentration of nitrogen at the plateau (for 80% N_2 gas) or shoulder (for 25% N_2 gas) was approximately 9wt.% N which corresponds to $(Fe,Cr)_{2.5}N$. This composition suggests the presence of ε nitride; this nitride was also determined by X-ray diffraction in the case of the 80% N_2 gas, but it was not detected in the case of the 25% N_2 gas[15]. In the diffusion zones in the Cr3Mo steel there was about 1.0 to 1.5% N.

3.2 Anodic Polarization

Figure 4 shows anodic polarisation curves for the outer surfaces of unnitrided and nitrided materials. Anodic currents for nitrided Armco iron were significantly lower than for untreated iron, being nearly the same for all the treatments. This demonstrates a good corrosion resistance of the compound zones. For Cr3Mo, the anodic currents after plasma nitriding were similar as for nitrided Armco iron, however, gas nitriding in ammonia did not appear to give a corrosion protection. Possibly, the lack of the protection was due to a very small thickness of the compound layer: from the GDS profile in Figure 3 it is apparent that the high-nitrogen superficial layer was only about 2μm thick. A lack of the improvement after the presently applied treatment of Cr3Mo can be ascribed to a high porosity of a thin compound layer, because thick ε nitride

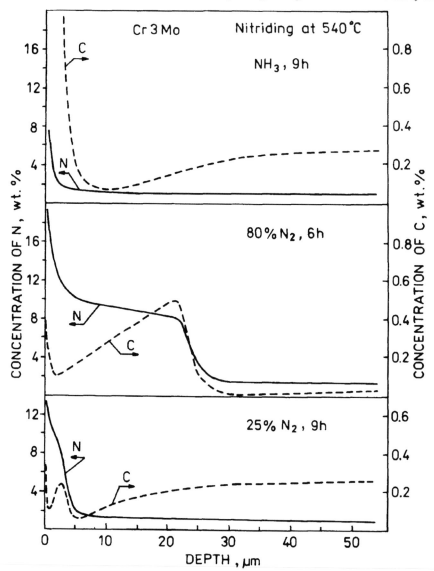

Figure 3 *GDS depth profiles for nitrogen and carbon in ammonia and plasma nitrided Cr3Mo*

layers obtained by conventional ammonia nitriding at 550°C for 64 h give a substantial improvement of corrosion resistance of iron and Fe-Cr alloys[14].

Anodic currents at various depths of plasma nitrided layers are shown in Figure 5 and Figure 6 for Armco iron and Cr3Mo steel, respectively. In Armco iron, an abrasion to the depths within the compound zone did not much affect the currents; evidently, the currents represent anodic behaviour of the iron nitride. These currents were similar for the low and high nitrogen atmospheres, so it was not possible to see a difference between the ε and γ'

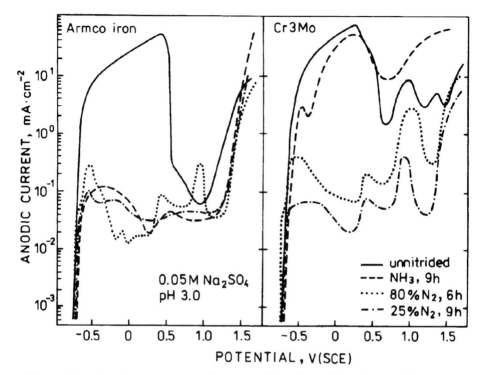

Figure 4 *Anodic polarization curves for unnitrided and nitrided surfaces of Armco iron and Cr3Mo steel in 0.05 M Na$_2$SO$_4$ at pH 3.0*

nitrides. At the depths of the transition from the compound to diffusion zones the currents were between those for the compound zone and for unnitrided iron. It appeared that the nitride precipitates did not significantly accelerate corrosion of the nitrided iron.

In the case of Cr3Mo steel (Figure 6), an abrasion within the compound zones resulted in a decrease in anodic currents. One of the reasons for this decrease can be a smoothing, and consequently a lowering of the true surface area. On the surfaces of the diffusion zones anodic currents were close to those for the unnitrided surface, or they exceeded the latter at the potentials before the transpassive region. Diffusion zones underwent general dissolution or pitting, depending on the depth. Pitting occurred at depths of about 15–25μm, whereas a strong dissolution with the formation of tiny pits was observed for deeper regions[16].

3.3 Depth Profiles of Hardness, Concentration, and Current

Figure 7 shows depth profiles of microhardness, of the nitrogen and carbon concentration, and of anodic current at the potential of 0.3 V (SCE) for Cr3Mo after plasma nitriding in the high-nitrogen atmosphere. The current at the potential of 0.3 V (SCE) corresponds to the minimum in the passive region (Figures 4–6), and therefore it can represent corrosion rate in the passive state.

Figure 7 indicates that the hardness changed with the depth more smoothly than the nitrogen content. Obviously, the high hardness in the diffusion zone is due to the nitride precipitates.

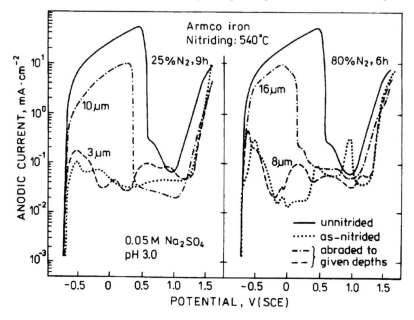

Figure 5 *Anodic polarization curves for unnitrided and plasma nitrided Armco iron in 0.05 M Na₂SO₄ at pH 3.0 after abrasion to given depths*

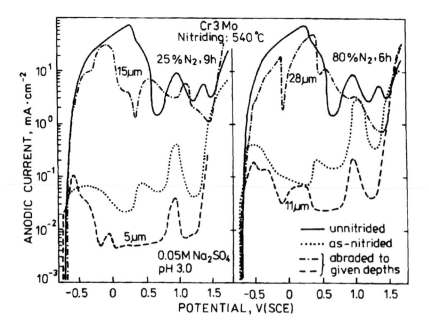

Figure 6 *Anodic polarization curves for unnitrided and plasma nitrided Cr3Mo in 0.05 M Na₂SO₄ at pH 3.0 after abrasion to given depths*

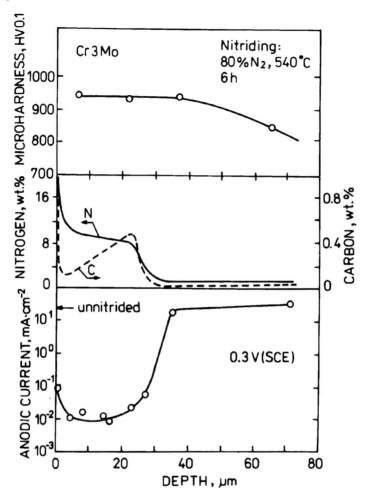

Figure 7 *Depth profiles of microhardness, of nitrogen and carbon concentration, and of anodic current at the potential of 0.3V (SCE) for Cr3Mo plasma nitride at 540°C for 6h at 80% N_2 + H_2*

Unlike the hardness profile, the anodic current profile closely corresponded to the nitrogen concentration. The currents were low for the nitrogen concentration of about 9% (ε nitride of $(Fe,Cr)_{2.5}N$), whereas they increased by about three orders of magnitude as the nitrogen concentration dropped to 1.5 %. The current in the diffusion zone was nearly the same as for the untreated steel, indicating a lack of a significant effect of the nitride precipitates on corrosion in the passive state.

4 CONCLUSIONS

Compound layers of ε nitride were formed on Cr3Mo steel during plasma nitriding at 540°C in the N_2/H_2 gas with 80% N_2 or 25% N_2, whereas on iron this nitride was formed only in the gas with 80% N_2. The γ' nitride formed on iron in the gas with 25% N_2. The gas nitriding used in this work did not produce on Cr3Mo a noticeable compound layer .

Compound zones exhibited significantly higher corrosion resistance than untreated materials. The resistance of these zones was for both materials similar. Diffusion zones showed poor corrosion resistance, being comparable with those of untreated surfaces. At shallow depths the diffusion zones in Cr3Mo underwent pitting corrosion, while in deeper regions they intensely dissolved with pin point etching.

There was a strict relationship between passive currents and the nitrogen concentration in nitrided layers. In the nitrided Cr3Mo steel only the compound zone (ε nitride) showed a high ability to passivate.

References

1. T. Bell, "Survey of the Heat Treatment of Engineering Components", The Metal Society, London, 1976.
2. H. C. Child, "Surface Hardening of Steel", Oxford University Press, Oxford, 1985.
3. "Ion Nitriding and Ion Carburizing" (Eds. T. Spalvins and W. L. Kovacs). ASM International, Materials Park, OH, 1990.
4. E. Haruman, T. Bell, and Y. Sun, *Surf. Eng.*, 1992, **8**, 275.
5. P. Sury, *Brit. Corros. J.*, 1978, **13**, 31.
6. K.-T. Rie and Th. Lampe, *Werk. Korros.*, 1982, **33**, 647.
7. E. Ebersbach, F. Henny, U. Winckler, G. Reisse, and C. Weissmantel, *Thin Solid Films*, 1984, **112**, 29.
8. U. Kamachi Mudali, B. Reynders, and M. Stratmann, *Materials Science Forum*, 1995, **185-188**, 723.
9. E. Angelini, B. De Benedetti and F. Zucchi, *La Metallurgica Italiana*, 1984, 499.
10. M. Jurcik-Rajman and S. Veprek, *Surf. Sci.*, 1987, **189/190**, 221.
11. U. Ebersbach, S. Friedrich, T. Nghia and H. J. Spies, *Härterei-Tech. Mitt.*, 1991, **46**, 339.
12. U. Ebersbach, S. Friedrich and T. Nghia, *Materials Science Forum*, 1995, **185-188**, 713.
13. A. E. Vorhoshkov and I. E. Sedloyev, Proceedings of 8th International Congress on Metallic Corrosion, Mainz, 1981; Dechema, Frankfurt am Main, 1981, Vol. 2, p. 1086.
14. K. Ibendorf and W. Schröter, *Neue Hütte*, 1986, **31**, 333.
15. J. Flis, J. Mankowski and E. Rolinski, in "Plasma Surface Engineering" (Eds. E. Broszeit, W. D. Munz, H. Oechsner, K.-T. Rie, and G. K. Wolf), DGM Informationsgesellschaft, Oberursel, 1989, Vol. **2**, p. 1165.
16. J. Mankowski and J. Flis, *Corros. Sci.*, 1993, **35**, 111.

Section 1.3 Wear

1.3.1

Modelling the Beneficial Effects of Oxidation on the Sliding Wear of Metals and Alloys

F. H. Stott[1], J. Jiang[2] and M. M. Stack[1]

[1]CORROSION AND PROTECTION CENTRE, UNIVERSITY OF MANCHESTER INSTITUTE OF SCIENCE AND TECHNOLOGY, MANCHESTER, UK

[2]PRESENT ADDRESS: DEPARTMENT OF AERONAUTICAL, MECHANICAL AND MANUFACTURING ENGINEERING, UNIVERSITY OF SALFORD, UK

1 INTRODUCTION

Damage to alloy surfaces during like-on-like sliding often involves an initial period of severe wear followed by a change to a lower wear rate or mild wear[1]. At the same time, the coefficient of friction decreases to a significantly reduced value. Although such changes can occur at low ambient temperatures, they are more pronounced at high temperatures; for many metals and alloys, there is a transition temperature, above which the wear rate in the mild-wear region becomes very low[2-4]. The main reason for the transition temperature is the establishment of smooth 'glaze' layers which give very good protection against wear and reduced resistance to sliding. It has been reported that the change from severe to mild wear is favoured by high oxidation rates of the metal[3,5], low load and low sliding velocities[7].

It has been proposed[2,8,9], and supported by experimental observations, that wear-debris particles play an important role in the establishment of the wear-protective layers accompanying the transition from severe to mild wear, even at low temperatures. For instance, artificially-supplied particles between sliding surfaces[10] and the application of magnetic fields, which help to retain magnetic particles within sliding surfaces[11], decrease significantly the time to attain mild-wear conditions. Conversely, removal of debris particles from the surfaces delays the change to mild wear[12]. Although pre-oxidation prior to sliding has little influence on the time to the onset of mild wear, prolonged exposure under static conditions at elevated temperatures, after some time of sliding, helps compaction of wear-debris particles due to sintering and promotes the onset of mild wear on subsequent further sliding[13]. The present authors have identified two types of wear-protective layers during sliding at elevated temperatures, compact wear-debris particle layers and smooth 'glaze' layers[14].

In this paper, the development of wear-protective layers during like-on-like sliding of Nimonic 80A (nominal composition: 18 to 20 wt% Cr, 1.8 to 2.7% Ti, 1.0 to 1.8% Al, balance Ni) at 20° to 600°C is reported and a model is described to account for the transitions from severe to mild wear.

2 EXPERIMENTAL RESULTS

Sliding-wear tests were undertaken in a pin-on-disc reciprocating rig at an average speed of 83 mm s[-1]. The pin specimen, with a 12.5 mm domed end, was placed in contact with a flat

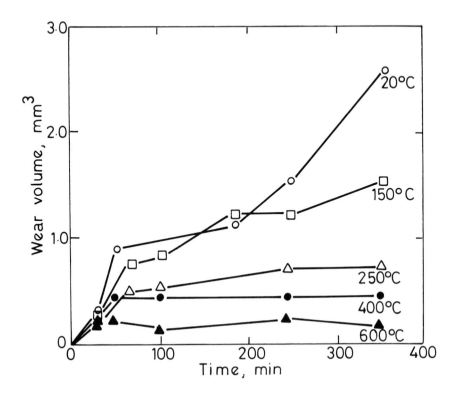

Figure 1 *Plots of wear volume of the pin versus time for Nimonic 80A during like on like sliding (load = 15N; average speed = 8.3 mms⁻¹) at various temperatures*

disc specimen, under a normal load of 15N in an environment of dry oxygen. Tests were carried out at ambient temperatures of 20°, 150°, 400° and 600°C. The friction force and electrical contact resistance between the sliding surfaces were monitored continuously. Wear was determined at the end of the test; the damage profiles were measured with a Talysurf attached to a microcomputer and the wear volumes were calculated by integrating the difference between the worn surface and the original surface. At the same time, a metal stylus and an electrical contact-resistance measurement system were used to monitor the distribution of high-resistance areas and their coverage on the wear scars; this gave an indication of the presence of oxide-rich wear-protective layers. Further experimental details are given elsewhere[14].

Figure 1 is a plot of wear volume of the pin specimen as a function of time for the five temperatures. Each point is from a different test. At 250°, 400° and 600°C, there were sharp transitions from high to low wear rates after short times of sliding; at 20° and 150°C, there were also changes to lower wear rates, although these were less marked than at the higher temperatures. The onset of the mild-wear regimes was associated with the development of high-resistance, oxide-containing, wear-protective layers, as illustrated by the estimated areas of the wear scars covered by such layers (Figure 2). These high-resistance regions tended to occur mainly near the centre and towards the ends of the wear scars where wear-debris particles accumulated. Changes in contact resistance from near zero to positive values and significant

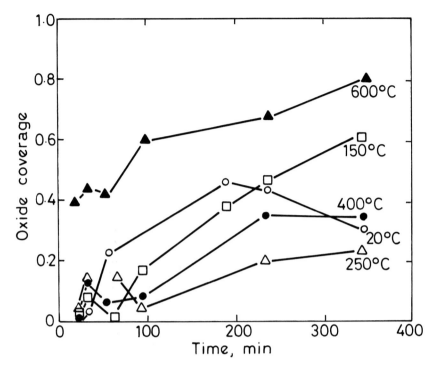

Figure 2 *Plots of oxide coverage on the wear scars versus time for Nimonic 80A during like-on-like sliding at various temperatures*

changes in coefficients of friction were also observed at the times of the onset of mild wear[14].

Examinations of the wear scars after the tests showed that two types of wear-protective layers had developed, depending on whether or not the temperature was below the critical value, apparently about 250°C under the present conditions. After sliding at the lower temperatures, many load-bearing areas were microscopically rough and covered by discrete but compacted particles of debris (Figures 3(a) and (b)). At the higher temperatures, very smooth 'glaze' layers were observed (Figures 3(c) to (f)). These had developed on and from compacted particles. Both types of layer were observed at temperatures above the transition, 250°C.

3 DISCUSSION

3.1 Development of Wear-Protective Surfaces During Sliding

Most theories of wear are concerned with generation of wear-debris particles which are then removed from the wear track, causing material loss. However, under the present conditions, although some debris particles are removed, others can become entrapped in the wear track and influence the subsequent sliding process. From experimental observations, the sliding wear process can be described as illustrated schematically in Figure 4. Wear debris particles are generated by the relative motion of the metallic surfaces under load. Some of these are removed from the wear tracks, resulting in metal loss. The others are retained within the

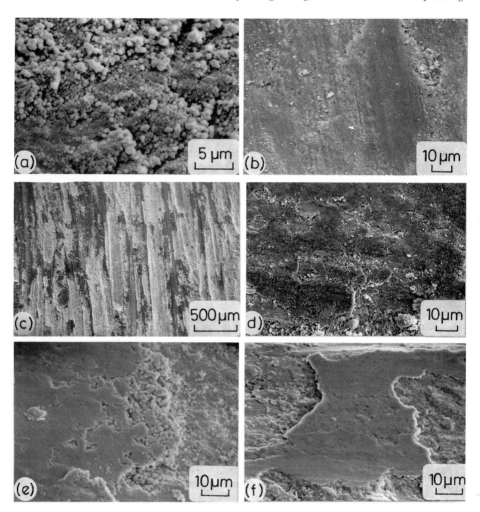

Figure 3 *Scanning electron micrographs of surfaces of Nimonic 80A after like-on-like sliding for 6 h at various temperatures: (a) 20°C, (b) 150°C, (c) 250°C, (d) 250°C, (e) 400°C and (f) 400°C*

tracks where they are comminuted by repeated plastic deformation and fracture while moving freely between the sliding surfaces. Once they have been reduced to a small enough size, they are agglomerated at certain locations on the contacting surfaces, due to the adhesion forces between solid surfaces arising from surface energy (15), and establish compact layers. This reduces material loss since newly-formed wear-debris particles are re-cycled into the layers; also, as the particles have been heavily deformed and oxidized, the layers are hard and wear-protective. On further sliding, two competitive processes take place: break-down of the layers, leading to wear (Figure 5(a)), and sintering/cold welding between the particles which consolidate the layer (Figure 5(b)). This process, together with oxidation of the particles, occurs more rapidly as the temperature is increased. If the surfaces of the layers become solid

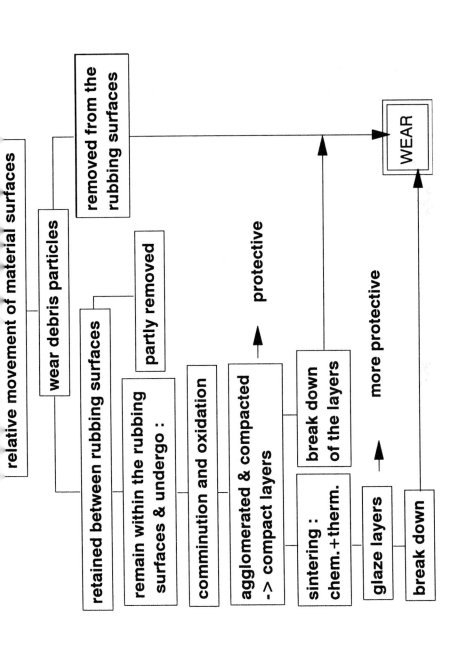

Figure 4 *Schematic diagram showing the sliding-wear processes at various temperatures*

Figure 5 *Scanning electron micrographs of surfaces of Nimonic 80A after like-on-like sliding for 6 h at 600°C: (a) Showing 'glaze' region breaking up, (b) Detail of edge of 'glaze' region*

before the layers are broken down, 'glaze' surfaces may develop on top of the compacted particle layers, leading to reduced friction and wear. The critical temperature is that temperature required to form a 'glaze' surface under a given set of conditions.

3.2 Model for the Establishment of Wear-Protective Layers

From the above description of the wear process, it is possible to develop a model for establishment of the wear-protective layers. It is assumed that the wear rate of the wear-protective layers is negligible compared to that of other areas and, hence, wear-debris particles are generated only in those other areas. Hence, the wear volume of the specimens after sliding time, t, is given by:

$$V(t) = \frac{\pi}{6} \int_o^t \{A(t) \, N(t) \, [1 - C_e(t)] \int_o^\infty [D^3 \, f(D) \, P_r(D)] \, dD\} \, dt \qquad \{1\}$$

where $N(t)$ is the number of wear-debris particles formed per unit time at time, t.

$A(t)$ is the apparent area of the wear scar at time, t

$C_e(t)$ is the effective coverage by wear-protective layers at time, t

$f(D).dD$ is the percentage of newly-generated particles that fall in the diameter range of D to $(D + dD)$.

$P_r(D)$ is the probability that a wear particle of diameter, D, is removed from the wear track.

The volume of wear-debris particles retained within the rubbing surfaces at time, t, is given by:

$$V_r(t) = \frac{\pi}{6} \int_o^t \{A(t)N(t)[1 - C_e(t)] \int_o^\infty D^3 f(D)[1 - P_r(D)] dD\} dt \qquad \{2\}$$

If it is assumed that all the retained particles are compacted eventually to form wear-protective layers, of average thickness, δ, which is typically 2 to 15 μm^{16}, then the coverage of the wear surface by such layers is given by:

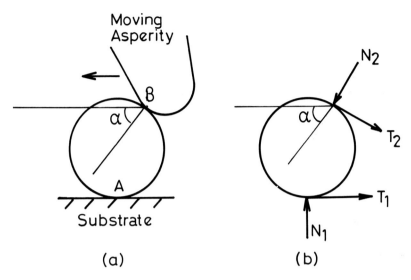

Figure 6 *Schematic diagram illustrating the contact conditions for a spherical debris particle between a plane surface and a protruding asperity from the countersurface*

$$C_{comp} = \frac{V_r(t)}{2A\delta} \qquad \{3\}$$

Wear-protective layers are particularly effective if 'glaze' oxide of a critical thickness, δ_g, can develop on their surfaces. If the oxidation kinetics of the wear debris are expressed in terms of the scale thickness, x, as

$$x = g(t)$$

then the critical time, t_c, for a glaze of critical thickness, δ_g, to develop on the compacted layer is given by:

$$t_c = g^{-1}(\delta_g) \qquad \{4\}$$

where $g^{-1}(\delta_g)$ is the inverse function of $g(t_c)$.

If an area of compact particle layer, $dA_c(\tau)$ is developed at time, τ, then an area of glaze equal to $dA_c(\tau)$ will have developed after a period of t_c, i.e. at time $(\tau + t_c)$. Thus, the total area of 'glaze' surface at time, t, is given by:

$$A_g(t) = \int_o^{t-t_c} dA_c(\tau) \qquad \{5\}$$

Hence, both the compact particle layers and the 'glaze' layers can co-exist on wear surfaces. They are both wear-protective, although the latter are more effective. From Figures 1 and 2, it is apparent that the coverage by 'glaze' layer (at 250° and 400°C) needed to cause the severe to mild wear transition is much less than that by the non-'glaze' compact particle layer (at 20° and 150°C). When both types of layer are present, the concept of an equivalent

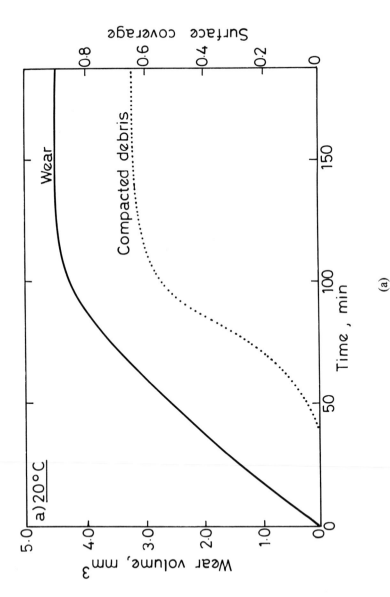

Figure 7 *Calculated plots of wear against sliding time under a load of 15N at (a) 20°C and (b) 400° and 600°C, showing the surface coverage by compacted debris and by 'glaze' layers*

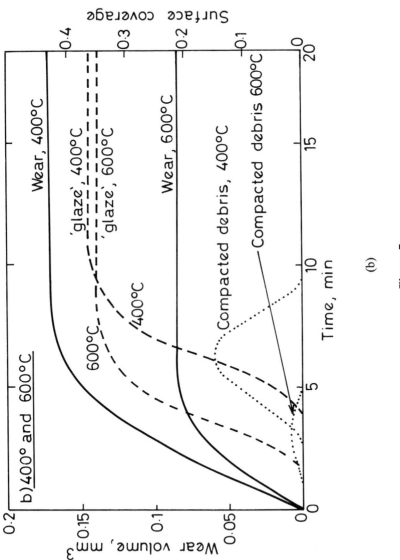

(b)

Figure 7

protective coverage of the wear surface can be introduced, defined as:

$$C_e = \frac{C_{comp}}{C_{climit}} + \frac{C_g}{C_{glimit}} \qquad \{6\}$$

where C_g is the surface coverage by 'glaze' layer.

C_{climit} and C_{glimit} are the critical coverage by compact and 'glaze' layers respectively above which the wear surfaces become completely protected and the transition from severe to mild wear occurs.

3.3 The Criteria for Removal of a Wear Debris Particle From the Sliding Surfaces

When a protruding asperity from a sliding surface comes into contact with a loose wear debris particle on the counter surface, the particle may roll forward, skid at the point of contact or become entrapped in the contacting surfaces; the first two result in removal of the particle from the surfaces. Here, it is assumed that the particle is a sphere of diameter, D. (Figure 6(a)). The static equilibrium conditions for the particle are indicated in Figure 6(b) where T_1, T_2, N_1 and N_2 are the local contact forces. It can be shown[17] that the critical criterion for the particle to be removed by either rolling or skidding occurs when:

$$\sin \alpha < \frac{1-f^2}{1+f^2} \qquad \{7\}$$

where $f = \min (f_1, f_2)$
$T_{1max} = f_1 N_1$
$T_{2max} = f_2 N_2$
T_{1max} and T_{2max} are the maximum frictional forces that can be attained.

Thus, particles that satisfy this criterion at their formation will move with the opposite surface. However, thereafter, they may be removed from the wear track or may become trapped at some location where the contact conditions are no longer favourable for movement. Conversely, particles that are trapped initially may be removed later if they are unable to agglomerate into a particle layer. For the purposes of this analysis, it is reasonable to assume that a particle is removed under two conditions: if it is not able to become incorporated into a compacted particle layer (compaction criterion) and if it satisfies the removal criterion of equation {7}, even when it remains at the lowest point on the wear surface (contact-geometry criterion). These two conditions have been considered separately, giving the following[17]:

i. The compaction criterion for a particle removal, P_1, is:

$$P_1 = \begin{cases} 1, & when\ D \geq D_{c1} \\ 0, & when\ D < D_{c1} \end{cases} \qquad \{8\}$$

$$\text{where } D_{cl} = \frac{3\pi}{8} \frac{\gamma \phi A}{W(P_a/N_p)_{crit}}$$

where

$$(P_a/N_p)_{crit} = \frac{1-k}{1-k^n}\left(\frac{1}{c}\frac{c+fs}{s-fc} - \frac{1}{s}\right)$$

$$k = \frac{s+f(1-c)}{s-f(1+c)}$$

$(P_a/N_p)_{crit}$ is the critical value of the relative adhesion force for a stable particle layer to form.

W is the total normal load

γ is the energy of adhesion between particles orginating from surface energy of the solid particles

ϕ is the ratio of real area of contact to apparent area of contact ($\sim$0.1)

$s = \sin \alpha$

$c = \cos \alpha$

ii. The compact-geometry criterion, P_2, is obtained from equation {7} by applying the Greenwood-Williamson surface contact model[18]. Here, it is assumed that a rough surface with hemispherical asperities is in contact with a rigid smooth counter surface. The height distribution of asperities is Gaussian, with 95% being higher than $Z = -2\sigma$; this height is taken as the lowest on the surface. From the analysis[17]:

$$P_2 = \begin{cases} 1 \text{ when } D \geq D_{c2} \\ 0 \text{ when } D < D_{c2} \end{cases}$$

{9}

where $\qquad D_{c2} = (1 + f^2)(d - 2\sigma)$

d is the separation between the two sliding surfaces.

σ is the standard deviation of the asperity-height distribution.

According to the surface-contact model[18], the surface separation, $d = h\sigma$, satisfies the following condition if plastic deformation of some asperities is involved[19]:

$$p = \frac{w}{AH\pi\eta\beta\sigma}$$

$$= \frac{4}{3\pi}\psi\int_h^{h+\frac{1}{\psi^2}}(s-h)^{3/2}\phi(s)ds + 2\int_{h+\frac{1}{\psi^2}}^{\infty}(s-h)\phi(s)ds$$

{10}

$$\text{where } \psi = \frac{E'}{H}\sqrt{\frac{\sigma}{\beta}}$$

and $\qquad$ p is the normalized load acting on two contacting wear surfaces.

η is the number of asperities per unit surface area.

β is the spherical tip radius of asperities on the surface.

E^1 is the elastic modulus of the contacting surfaces.

H is the hardness of the contacting surfaces.

Overall, the probability of particle removal, P_r, is obtained from the above two probabilities, P_1 and P_2:

$$P_r \;=\; P_1 \cup P_2 \;=\; \begin{cases} 1, \text{ when } D \geq D_c \\ 0, \text{ when } D < D_c \end{cases}$$

where D_c is min (D_{c1}, D_{c2})

3.4 Validity of the Model

Using the model, the wear rates expected from the tests in the reciprocating sliding rig have been calculated and compared with those obtained in the tests. Most data used in the calculations have assumed that the specimens were nickel since those for the alloy are less readily available. The rate of generation of wear-debris particles and their average size during sliding of nickel have been determined experimentally[20]. The same surface contact parameters as assumed by Greenwood and Williamson[18] have been used, namely $\eta = 300/\text{mm}^2$, $\beta\sigma = 10^{-4}$ mm^2, $E^1(a/b)^{\frac{1}{2}} = 25$ kg/mm^2; the hardness of the alloy was taken as 500 kg/mm^2. It was assumed that growth of nickel oxide films on nickel follows a logarithmic relationship with time at temperatures below 350°C and for films < 3 nm thick while, for films thicker than 3 nm and at temperatures of 300° to 700°C, growth follows a parabolic relationship with time. The experimental data obtained for oxidation of nickel by Graham and Cohen[21] were used in the calculations. Although the eventual steady-state oxide on Nimonic 80A is not nickel oxide, growth of this oxide is likely to predominate in the early stages, especially at low temperatures. Yamamoto and Nakajima[22] have determined the values of the adhesion energy, γ, as a function of temperature for copper in contact with chromium and for the chromium in contact with mica. As these values should not be much different from those of nickel-base alloys, they have been used in the calculations. Finally, it was assumed that the size distribution of wear-debris particles when they were just generated, $f(D)$, follows a Gaussian distribution:

$$f(D) \;=\; \frac{1}{\sqrt{2\pi}\,\sigma_D}\,\exp\left(-\frac{(D-D_o)^2}{2\sigma_D^2}\right) \qquad \{11\}$$

where the standard deviation of particle size, σ_D, was estimated according to the following relationship so that 99% of the particles were within the range of $(D_o - 1/3D_o)$ to $(D_o + 1/3D_o)$, based on the fact that the minimum size of the debris particles is generally 1/3 of the average size[23]:

$$3\sigma_D = D_o - 1/3D_o \qquad \{12\}$$

Figure 7 shows the calculated variations of wear volume and surface coverage by compacted

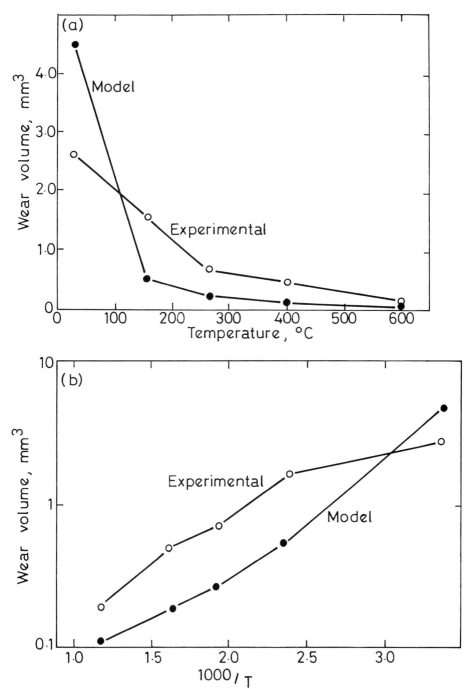

Figure 8 *Comparison of the calculated and experimentally-observed variations of wear volume (after 6 h sliding) with temperature under a load of 15N, (b) is a replot of (a) according to the Arrhenius relationship*

debris particles and by 'glaze' layers as functions of sliding time. In the early stages, metal-metal contact occurs and the large contact stresses, arising from the specimen geometry, produce large wear-debris particles; as indicated by equations[8-10], large particles and high contact pressures result in a high probability of removal of particles from between the sliding surfaces. Thus, compacted wear-protective layers do not develop in the early stages. As sliding continues, the size of the wear scar increases and the contact pressures are reduced, while the wear-debris particles entrapped between the sliding surfaces are comminuted, producing particles that are eventually smaller than the critical size (equation {8}). Thereafter, the surface coverage by wear-protective compact particle layers increases rapidly, causing a sharp decrease in wear rate (Figure 7(a)). This is consistent with experimental observations that such layers develop quickly at about the time of the friction and wear transitions[2]. At higher temperatures, 'glaze' layers develop on top of the compacted debris as a result of increased sintering and oxidation of the particles, resulting in more solid and wear-protective surfaces (Figure 7(b)).

3.5 The Effects of Temperature on Wear

Figure 8(a) compares the calculated wear volumes with the experimentally-observed volumes after 6h sliding. In both cases, the wear decreases rapidly with increasing temperature at about 250°C. The reasons for this are not immediately clear since the development of smooth 'glaze' layers due to increased oxidation on increasing temperature cannot account fully for the observations. This is apparent if the data are re-plotted according to an Arrhenius relationship (Figure 8(b)). The activation energy for the wear process, Q, is approximately 3000 cal/mole; this is more than one order of magnitude less than that for oxidation of a metal but is within the range of activation energy for physical adsorption interactions between solid surfaces[22]. It is also within the range of activation energy for the development of adhesion between contacting asperities in the stick-slip static-friction process[24]. Although triboxidation is different from normal oxidation, the activation energy is essentially the energy barrier for the kinetic processes occurring within the oxide layers and should be similar to that for normal oxidation[16]. Thus, the low activation energy for the wear process suggests that the development of wear-protective layers is not controlled by the oxidation process but is closely related to adhesion between wear-debris particles in the wear surfaces, consistent with the present model. Hence, the major effect of increased temperature is that more wear-debris particles are involved in the establishment of the compact wear-protective layers.

4 CONCLUSIONS

1. There is a transition from severe to mild wear during like-on-like reciprocating sliding of Nimonic 80A in oxygen at 20° to 600°C; this is accompanied by the development of wear-particles layers, having high electrical resistance, on the contacting surfaces.
2. At 20° and 150°C, the wear-protective layers consist essentially of compacted wear-debris particles; they are developed from agglomeration and compaction of fine wear debris entrapped within the sliding surfaces. Some wear continues on further sliding after the onset of mild wear. At temperatures above about 250°C, smooth 'glaze' layers develop on these compacted particulate regions and the wear rate in the mild-wear regime is negligibly small.

3. The development of wear-protective layers and the transitions from mild to severe wear can be described by a mathematical model that is based on the probabilities of retention or removal of wear-debris particles from the sliding surfaces. Reasonable agreement has been achieved between calculated results and experimental observations at temperatures from 20° to 600°C.

Acknowledgements

The authors are grateful to the British Council for support of Jiaren Jiang under the Sino-British Friendship Scholarship Scheme.

References

1. J. F. Archard and W Hirst, Proc Roy Soc London, 1956, **A236**, 397.
2. F. H. Stott, J. Glascott and G. C. Wood, *Wear*, 1985, **101**, 311.
3. A. F. Smith, *Tribology International*, 1986, **19**, 65.
4. P. T. Newman and J. Skinner, *Wear*, 1986, **112**, 291.
5. T. S. Eyre and D Maynard, Wear, 1971, **18**, 301.
6. N. C. Welsh, Phil Trans Roy Soc London, 1965, **A257**, 31.
7. J. K. Lancaster, Proc Roy Soc London, 1963, **A273**, 466.
8. F. H. Stott and G. C. Wood, High Temperature Corrosion, NACE 6, NACE, Houston, 1983, p. 406.
9. F. H. Stott, J. Glascott and G. C. Wood, Corrosion-Erosion of Materials at Elevated Temperatures, ed A V Levy, NACE, Houston, 1987, p. 263.
10. A. Iwabuchi, *Wear*, 1991, **151**, 301.
11. K. Hiratsuka, T. Sasada and S. Norose, *Wear*, 1986, **110**, 251.
12. E. R. Leheup and R. E. Pendlebury, *Wear*, 1991, **142**, 351.
13. J. L. Sullivan and N. W. Granville, *Tribology International*, 1984, **17**, 63.
14. J. Jiang, F. H. Stott and M. M. Stack, *Wear*, 1995, **176**, 185.
15. K. L. Johnson, K. Kendall and A. D. Roberts, *Proc Roy Soc London*, 1971, **A324**, 301.
16. T. F. J. Quinn, *Wear*, 1984, **94**, 175.
17. J. Jiang, F. H .Stott and M. M. Stack, *Wear*, 1995, **181–183**, 20.
18. J. A. Greenwood and J. B. P. Williamson, *Proc Roy Soc London*, 1966, **A295**, 300.
19. K. A. Nuri and J. Halling, *Wear*, 1993, **160**, 213.
20. N. Soda, Y. Kimura and A. Tanaka, *Wear*, 1975, **35**, 331.
21. M. J. Graham and M. Cohen, *J. Electrochem Soc.*, 1972, **119**, 879.
22. N. Yamamoto and K. Nakajima, *Wear*, 1981, **70**, 321.
23. E. Rabinowicz, *Wear*, 1964, **7**, 9.
24. C. A. Brockley and H. R. Davis, *Trans ASME*, 1968, **F90**, 35.

1.3.2

Oxidation and Erosion Resistance of Amorphous Bright Chromium Electrodeposited Coatings

R. P. Baron and A. R. Marder

ENERGY RESEARCH CENTER AND DEPARTMENT OF MATERIALS SCIENCE AND ENGINEERING, LEHIGH UNIVERSITY, BETHLEHEM, PA 18015, USA

1 INTRODUCTION

One of the main causes of external coating failure is high temperature corrosion damage. A popular method of combating this problem is applying a diffusion coating by chromizing, a pack surface treatment process at elevated temperatures in which an alloy is formed by the inward diffusion of chromium into the base metal[1]. Chromium has excellent resistance to wear, heat and many corrosive media. However, the chromizing process causes porosity, voids, colunmar grains and grain boundary carbides in the microstructure which can greatly inhibit the coating performance. Earlier studies have indicated that the detrimental voids and carbides were formed during the slow furnace cooling of the chromizing process[2].

Electrodeposition is a method of chromium deposition that shows a great deal of promise for applying chromium without the harmful effects incurred by the chromizing process. However, upon plating, cracks can develop that act as easy paths to the base metal for corrosion products. To prevent this corrosion attack, a crack-free chromium plate can be produced by manipulating the plating conditions[3,4]. Thus, the purpose of this research is to successfully produce a crack-free chromium coating and then to test the microstructural integrity of this coating at elevated temperatures in an oxidizing environment and under erosion impact conditions. Along with manipulation of the elecrodeposition current, the composition of the electrolytic bath can also be altered to produce a crack-free plate, called the *amorphous bright chromium deposition method* (ABCD)[5,6].

2 EXPERIMENTAL PROCEDURE

The substrate material used was a high purity iron-carbon alloy containing 0.20 wt% C. The substrates were machined to rectangular coupons with dimensions 38x19x3mm and were then ground on 120 grit to remove any predominant burrs and oxide formation. After the cleaning, the samples were rinsed with deionized water and placed in the amorphous bright chromium deposition (ABCD) solution. The ABCD plating solution was contained in a pyrex beaker and had the following composition: 100g/l of chromic acid (CrO_3), 5g/l of sulphuric acid (H_2SO_4), and 20 ml/l of an 85% solution of formic acid[5,6]. To increase the current efficiency, the temperature of the plating bath was maintained around 20°C with an ice bath surrounding the beaker. The current density was set at 40 A/dm^2, following the published optimum value for the ABCD technique from Hoshino et al[5]. To maintain adequate circulation of the plating solution during deposition, the solution was mixed using a magnetic stirrer set at a speed of 300 rpm. The duration of deposition was 30 minutes. To remove any adherent solution after plating, the sample was rinsed with deionized water and wiped clean.

After plating, the samples were sectioned with a diamond saw to a size of 20 × 10mm, removing the unplated portion, and vacuum encapsulated. The samples were then placed in a muffle furnace at a temperature of 700±7°C. Five different exposure times were used: 5, 10, 20, 50 and 200 hours. When the samples were removed from the furnace, they were quenched in cool water in order to "freeze in" the microstructure produced at 700°C.

To study the high temperature corrosion behaviour of the ABCD coatings, plated samples were placed in the furnace without being encapsulated. These samples were subjected to an oxygen atmosphere at 700°C for various exposure times up to a maximum of 1200 hours. Unlike in the encapsulation studies, the plated samples for the oxidation studies were allowed to air cool from 700°C to preserve the surface oxidation and to prevent any cracking of the ABCD plate from thermal stress.

To determine the effect of an annealing step on the oxidation behaviour of the ABCD coating, as-plated samples were encapsulated and exposed at 700°C for various times. Two pieces of the same ABCD plate were placed in each encapsulation. The encapsulated exposure times were: 1, 5, 22, 200 and 500 hours. After exposure, the samples were quenched and ultrasonically cleaned in alcohol. For each encapsulation, one piece of the ABCD plate was saved to characterize the effects of the annealing step on the properties of the plate; the other piece was exposed to an oxidizing atmosphere at 700°C for 500 hours.

To investigate the applicability of the ABCD coating for erosive environments, room temperature erosion tests were performed on as-plated and annealed samples. The annealed samples were as-plated samples that were encapsulated, exposed for one hour at 700°C, and quenched in cool water. The samples were cut into 1.25 × 1.25mm squares. The angle of impingement was 90°, the erodent material was alumina, the erodent velocity was 40 m/s, and, finally, the feed rate was approximately 80 grams/minute. To obtain steady-state erosion, five test times were chosen: 10, 30, 50, 80 and 100 minutes. The erosion test apparatus is described in detail elsewhere[7].

3 RESULTS AND DISCUSSION

3.1 Heat Treatment

After exposure to 700°C for various times, all the vacuum encapsulated samples showed very little oxide formation, indicating that the vacuum imposed was sufficient to prevent atmospheric effects. The exposed samples had a greyish, matte appearance as opposed to the bright unexposed plated samples and the severity of cracking appeared to increase with exposure time. A cross-sectional micrograph of a 200 hour diffused coating is shown in Figure 1. From the micrograph five distinct regions can be observed, all separated by relatively planar interfaces. This type of alloy layer formation was observed in each encapsulated sample for all exposure times. Even though the as-plated ABCD coating contained relatively few cracks, the number of cracks did not increase greatly after the elevated temperature exposure. The exposed coating had approximately 1.6 cracks/mm, compared to 1.4 cracks/mm prior to exposure. Conventional chromium plating exhibits on the order of 100 cracks/mm.

To determine the behaviour of the structure of the ABCD plate after exposure at 700°C for various times, X-ray diffraction was performed on the annealed samples. The encapsulated samples were exposed for 1, 5, 22 and 200 hours and their diffraction patterns with peak labels are shown in Figure 2. Even after only one hour of exposure at 700°C, the original diffused chromium peak seen in the amorphous as-plated sample has transformed into a distinct chromium peak (at 44°). Other minor chromium peaks have also formed along with the development of chromium carbide and chromium oxide peaks. The pattern seems to strongly indicate the presence of Cr_7C_3 which can be attributed to the large amount of C from the formic acid (HCOOH) in the plating bath. This result is consistent with the diffraction patterns of Hoshino et al[5] and Tsai and Wu[6] after one hour exposure at 700°C.

As the exposure time is increased to 22 hours, a continued increase in the peak intensity of the chromium and Cr_7C_3 phases is observed. However, when the exposure time is increased to 200 hours, a distinct change in the diffraction pattern occurs. The crystalline chromium and chromium oxide peaks still exist, but now the previously numerous Cr_7C_3 peaks are no longer obvious. Some peaks exist which may indicate the presence of Cr_7C_3, but due to the oxide scale and errors involved in the analysis, no definitive identification can be made and a new type of chromium carbide, $Cr_{23}C_6$, dominates the diffraction pattern. Due to the large difference in exposure time between the 22 hour and 200 hour exposure samples, the exact exposure time at which the $Cr_{23}C_6$ phase forms could not be determined. Tsai and Wu[6] found similar results for increasing temperatures.

The room temperature microhardness was measured on the top surface of five plates after they had been exposed for 1, 5, 22, 200 and 500 hours respectively, and the Vickers hardness values are plotted as a function of exposure time in Figure 3. Included in Figure 3 is the Vickers hardness measured by Hoshino et al[5], after one hour's exposure at 700°C. The data show that when the ABCD plate is exposed for one hour, the hardness increases dramatically which can be related to the crystallization of the amorphous chromium and the precipitation of Cr_7C_3 from the amorphous matrix. However, as the exposure time is increased, the hardness of the coating decreases. Hoshino et al[5] and Tsai and Wu[6] demonstrated that the hardness of the ABCD plates reaches a maximum after one hour exposure at 500°C and then begins to diminish as the exposure temperature is increased. Tsai and Wu[6] attributed this decrease in

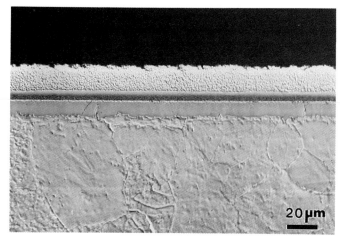

Figure 1 *Cross-sectional light optical micrograph of encapsulated sample after 200 hours of exposure*

hardness to increasing chromium dispersoids and chromium carbide size. Also, as the temperature is increased further, Cr_7C_3 is replaced by Cr_3C_6, a lower hardness carbide. (Cr_7C_3 typically has a hardness of 2100 DPH compared to 1650 DPH for Cr_3C_6.) Thus, the decrease in hardness with extended exposure time at 700°C can also be explained by continued growth of the chromium dispersoids and precipitation of the Cr_3C_6 phase. The longer the exposure time, the greater the coarsening of the crystalline chromium dispersoids and precipitation of the Cr_7C_6 phase and as a result, the lower the hardness value.

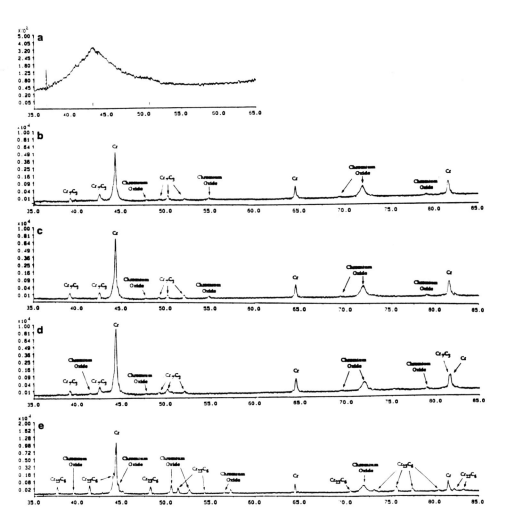

Figure 2 *X-ray diffraction patterns of ABCD chromium coating (a) as-plated and annealed at 700°C for (b) 1 hour, (c) 5 hours, (d) 22 hours and (e) 200 hours*

3.2 Elevated Temperature Oxidation Studies

Even though long cracks are visible in the plate before exposure, the extent of cracking and the width of the cracks appear to increase with increasing exposure. Figure 4 shows a representative microstructure of the exposed plates. The oxide first attacks down a pre-existing crack and then travels along the coating/substrate interface. The oxide also "mushrooms" up through the original crack. The lateral percentage of oxide attack versus exposure time was determined, Figure 5. The lateral percentage was calculated by measuring the planar length of oxide attack observed at the interface over the entire length of sample observed. At first, as the exposure time is increased, the lateral oxide attack at the interface resembles parabolic growth until 500 hours. After this exposure time, the lateral percentage of oxide attack for all the samples jumps over 20%, indicating "break-away" corrosion and another parabolic attack. Since carbide layers have good corrosion resistance[8], sufficient formation of such a layer before substantial oxide attack could increase the corrosion resistance. Therefore, the samples that form the carbide layer the quickest should show an increased resistance to oxide attack. With this in mind, ABCD plates on non-decarburized substrates were first encapsulated and then exposed to 700°C for various times, namely: 1, 5, 22, 200 and 500 hours. These samples were then exposed to an oxidizing atmosphere, again at 700°C for 500 hours. By observing the oxidation attack, the effectiveness of forming the alloy layer can be determined.

Figure 6 shows the lateral percentage of oxide attack as a function of pre-anneal exposure time. These results indicate that, within experimental error, there is little change in the oxide attack as the annealing time at 700°C is extended to five hours. However, when the exposure time reaches 22 hours, a significant decrease in the attack is observed, which can be accredited to the formation of carbide layers below the ABCD coating. Therefore, even though the

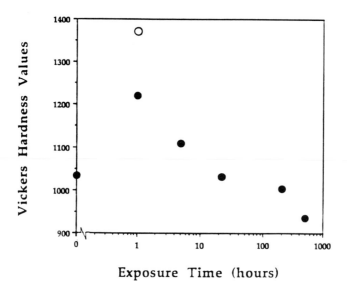

Figure 3 *Vickers hardness values as a function of exposure time. The temperature of exposure was 700°C. The open circle represents the value measured by Hoshino et al[5]*

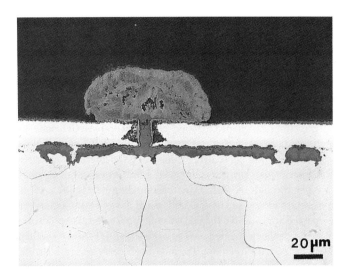

Figure 4 *Cross-sectional light optical micrograph of ABCD plate oxidized for 200 hours at 700°C*

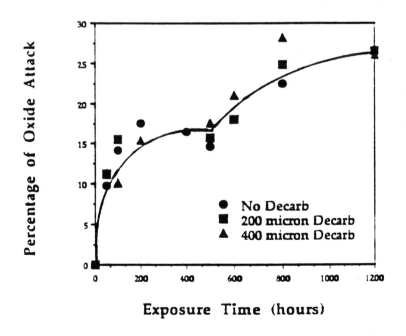

Figure 5 *Lateral percentage of oxide attack as a function of exposure time*

ABCD initially contains macrocracks, the formation of the alloy layers prevents accelerated oxide penetration down these cracks when the plate is exposed to elevated temperature oxidation. Figure 7 shows a pre-existing macrocrack in the ABCD coating in the 22 hour anneal sample after 500 hours of elevated temperature oxidation exposure. No extensive oxidation attack is visible.

When the pre-annealing step is increased to 200 hours, the oxide attack at the interface has a different appearance. Due to the extended time of the pre-anneal, the carbide layers have grown substantially and no visible attack is observed down any of the pre-existing cracks. Thus, instead of travelling down pre-existing cracks in the ABCD plate and forming a continuous layer, the oxide attack occurs in discrete areas along the coating/substrate interface. Figure 8 shows the formation of spherical cavities along the interface and internal oxidation attack[9]. The formation of these cavities at the coating/substrate interface may be caused by the diffusion of oxygen through the ABCD plate along discrete carbide particles.

Although the oxide attack, caused by the diffusion of oxygen through the coating, appears to be less severe than the attack down the pre-existing cracks, the planar percentage of attack becomes greater as the pre-anneal is increased past 22 hours (Figure 6). When the attack is down macrocracks, the penetration along the interface is very severe near the macrocrack, but other areas of the interface are untouched. On the other hand, when the attack is caused by the diffusion of oxygen through the coating, the discrete penetrations extend along the entire coating/substrate interface, thereby increasing the percentage of oxide attack.

3.3 Erosion Tests

The weight loss of the as-plated and annealed ABCD samples as a function of erosion exposure time is shown in Figure 9. To compare the erosion behavior of the as-plated and annealed plates, the data points can be fit with a straight line and the slope of this straight line yields the steady-state erosion rate. The steady-state erosion rate is an indication of the extended erosion behaviour of a material. For the as-plated ABCD samples the erosion rate is 0.080 mg/min. However, for the annealed ABCD samples, the steady-state erosion rate is 0.042 mg/min, considerably less than for the as-plated samples.

The significant increase in erosion resistance of the annealed ABCD plate can be attributed to the crystallization of the amorphous chromium and precipitation of chromium carbides. As previously exhibited, when the ABCD plate is annealed for one hour at 700°C, chromium crystallizes and the Cr_7C_3 phase precipitates out of the amorphous matrix. As a consequence, the hardness increases (from 1033 to 1221 HV), principally from the precipitation of the carbide phase. In effect, the annealed ABCD plate becomes a dispersion-hardened alloy – hard carbides in a relatively "soft" chromium matrix. Together, the hard chromium carbides and the "soft" chromium matrix may provide the optimal combination of mechanical properties for better erosion resistance compared to the as-plated, amorphous plate.

4 CONCLUSIONS

By exposing amorphous bright chromium deposits (ABCD) to 700°C for various times under neutral and oxidizing atmospheres, the heat treatment, corrosion and erosion behaviour of the electroplates were characterized. From these analyses, the following conclusions can be drawn:

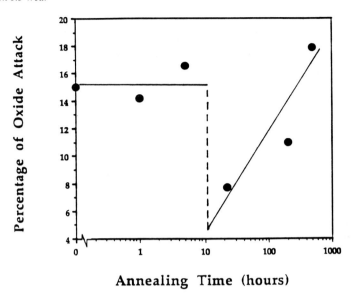

Figure 6 *Lateral percentage of oxide attack versus the pre-anneal exposure time at 700ºC. Each sample was exposed to an oxidizing atmosphere for 500 hours*

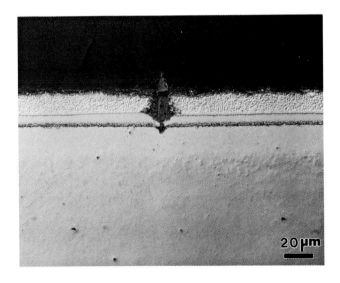

Figure 7 *Cross-sectional light optical micrograph of unaffected macrocrack in the ABCD plate after 22 hours anneal and 500 hours oxidation at 700ºC*

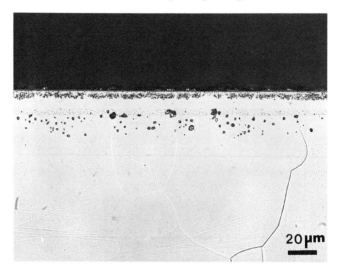

Figure 8 *Cross-sectional light optical micrograph of spherical penetration of oxide at coating/substrate interface in ABCD plate after 500 hour anneal and 500 hour oxidation at 700°C*

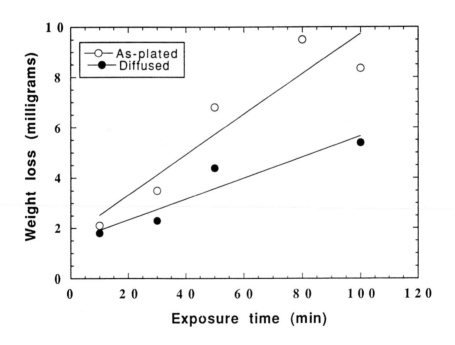

Figure 9 *Erosion weight loss exposure time for as-plated and annealed ABCD plates*

1. The ABCD technique utilized in the present study produced a uniform, bright amorphous chromium plate. Although macrocracks formed throughout the plate, the number of cracks in the ABCD plate is greatly reduced to approximately 1.4 cracks/mm compared to the conventional chromium plate which is on the order of 100 cracks/mm.

2. From the X-ray diffraction studies, when the ABCD plate is vacuum encapsulated and annealed at 700°C for one hour, chromium crystallizes and Cr_7C_3 precipitates out of the amorphous chromium matrix increasing the hardness of the coating. These constituents continue to grow as the exposure time is increased to 22 hours. However, after 200 hours of exposure at 700°C, the $Cr_{23}C_6$ phase completely replaces the Cr_7C_3 phase lowering the hardness of the coating.

3. A pre-anneal heat treatment to form alloy layers prior to the oxidation exposure of the ABCD plates showed a significant decrease in the planar oxide attack at the interface when the pre-anneal was extended to 22 hours. However, when the pre-anneal time was extended to 200 hours, the oxide attack increased and was attributed to discrete oxide particle formation at the coating/substrate interface.

4. Room temperature erosion behaviour of as-plated and annealed ABCD plates was studied. After a one hour anneal at 700°C, the steady-state erosion rate decreased from 0.080 mg/min compared to the as-plated result of 0.042 mg/min. This increase in erosion resistance was attributed to the precipitation of a hard Cr_7C_3 phase in a relatively soft chromium matrix, effectively making the annealed ABCD plate a dispersion-hardened alloy.

Acknowledgement

The authors would like to thank their colleagues A. O. Benscoter and J.N. DuPont for their help on this research. The invaluable assistance of D. Susan in preparing the manuscript and the sponsorship of the project by PSE & G, PP & L, APS, Virginia Power, and Ohio Edison is greatly appreciated.

References

1. American Society for Metals, 'Metals Handbook Desk Edition', 1985, p. 1.8.
2. V. Agarwal and A. R. Marder, *Microstructural Characterization*, 1996, **36**, 35.
3. B. A. Wilson and D. M. Turley, *Trans. of the Institute of Metal Finishing*, 1989, **67**, 104.
4. J. C. Saiddington and G. R. Hoey, *Plating*, 1974, **61**, 923.
5. S. Hoshino, H. Laitineu and G. B. Hoflund, *J Electrochemical Society*, 1986, **133**, 681.
6. R-Y. Tsai and S-T Wu, *J Electrochemical Society*, 1989, **136**, 1341.
7. B. A. Lindsley, K. Stein, and A. R. Marder, *Meas. Sci. Technol.*, 1995, **6**, 1169.
8. W. Kaluba and A. Wachowiak, *Arch. Eisenhuttenwes*, 1983, **54**, 161.
9. F. H. Stott and G. C. Wood, *Materials Science and Technology*, 1988, **4**, 1072.

1.3.3

The Effect of Chromium Carbide Coatings on the Abrasive Wear of AISI 44OC Martensitic Stainless Steel

A. B. Smith[1], A. Kempster[2], J. Smith[2] and R. Benham[1]

[1]INSTITUTE OF POLYMER TECHNOLOGY AND MATERIALS ENGINEERING, LOUGHBOROUGH UNIVERSITY OF TECHNOLOGY, LEICESTERSHIRE, LE1 1 3TU, UK

[2]DIFFUSION ALLOYS LIMITED, HATFIELD, HERTFORDSHIRE, AL9 5JW, UK

1 INTRODUCTION

Chromium carbide coatings are principally used where resistance to corrosion, in addition to wear resistance, is important[1]. They have, for many years, been deposited by the closed-reactor chemical vapour deposition (CVD) process of pack cementation (i.e. chromising)[2]. In this process, the parts to be coated are packed in and reacted with a powder mixture within an essentially closed container. This mixture consists of chromium, a halide activator and an inert diluent. Contact with the pack can, however, lead to the detrimental retention of small particles of pack material on coated surfaces, increasing surface roughness and often necessitating post-coating finishing[2,3]. For this, and other reasons, there has been a growing trend to the deposition of chromium carbide coatings by the conventional, open-reactor CVD process[1-3], in which the gaseous reactants are continuously supplied to the coating chamber and the by-product gases continually removed from it. For brevity, this process will hereafter be referred to simply as *gas CVD*. Recently, another alternative to the pack cementation process has been developed. It resembles pack cementation in the sense that a powder mixture is employed, but differs from it in that the parts to be coated with chromium carbide are placed out of but directly above this mixture[4], obviating the retention of powder particles on the coated surfaces. This process is termed *out-of-pack cementation*. It further differs from pack cementation in that it involves hydrogen and/or argon flow through the coating chamber, and may, consequently, be categorised as a semi-open-reactor CVD process.

Certain characteristics of chromium carbide coatings deposited by the pack and gas CVD processes have previously been reported to be basically equivalent[2,5]. However, although the wear retarding effect of both pack[6] and gas CVD[7] chromium carbide coatings has been referred to in the literature, no assessment of their relative wear resistance appears to have been made.

Recently, a detailed study was reported in which an overall comparison was made of the characteristics of chromium carbide coatings produced on AISI 44OC martensitic stainless steel substrates, not only by pack cementation and gas CVD, but also by out-of-pack cementation[8]. The purpose of the present work was to extend this study by comparing the effect of the three chromium carbide coatings on the abrasive wear of the AISI 44OC martensitic stainless steel substrates.

2 EXPERIMENTAL WORK AND RESULTS

2.1 Chromium Carbide Coated and Uncoated Samples

The AISI 44OC martensitic stainless steel samples (nominal composition, 0.95–1.20 wt.% C, 16.0–18.0 wt.% Cr, 0.75 wt.% max. Mo, bal. Fe) were initially milled from wrought bar and then ground to dimensions of 12.7 mm square by 3.18 mm thick. For chromium carbide coating by the pack cementation process, a number of these samples were packed in a powder mixture of ammonium iodide, chromium and inert filler and held at a temperature of 1020°C. In the case of the out-of-pack cementation process, the samples were placed directly above a mixture of ammonium chloride, alumina and chromium powder in a 5%H_2/Ar atmosphere at a temperature of 1050°C. Finally, the gas CVD chromium carbide coatings were deposited on the samples at atmospheric pressure and a temperature of 1000°C, using the standard reaction of $CrCl_2$ with H_2. The $CrCl_2$ reactant was generated in situ by the passage of HCl over chromium chips. No gaseous hydrocarbon reactant was used. The deposition time employed for each process was chosen with the aim of producing coatings of comparable thickness. No post-coating finishing of the pack coated samples was carried out. However, these and the out-of-pack and gas CVD coated samples were all vacuum heat treated to harden their stainless steel

Table 1 *Details of uncoated and chromium carbide coated abrasive wear test samples*

Sample number	Sample thickness (mm)	Sample (substrate) hardness (HV_{30})	Coating thickness (μm)
Uncoated samples			
Sample 1	3.17	639	-
Sample 2	3.17	652	-
Sample 3	3.17	639	-
Sample 4	3.18	652	-
Pack coated samples			
Sample 1	3.21	606	16.01
Sample 2	3.21	602	15.99
Sample 3	3.21	614	15.39
Sample 4	3.20	606	15.83
Out-of-pack coated samples			
Sample 1	3.20	583	4.42
Sample 2	3.17	594	4.81
Sample 3	3.17	575	6.55
Sample 4	3.18	586	5.56
Gas CVD coated samples			
Sample 1	3.18	610	8.06
Sample 2	3.18	602	8.34
Sample 3	3.18	618	8.03
Sample 4	3.18	618	8.03

substrate. The procedure employed was as follows: preheat at 775°C, austenitise at 1030°C, N_2 gas quench to room temperature and double temper at 200°C for one hour. The uncoated samples were heat treated in exactly the same way. It is pertinent to note that the deposition conditions (not including deposition time) and the heat treatment conditions detailed above were identical to those employed in the study of chromium carbide coating characteristics cited earlier[8]. This being the case, it is considered justifiable to assume that the chromium carbide coating characteristics determined in that study (with the exception of coating thickness) will be germane to the present work.

Following vacuum heat treatment, the thickness of the chromium carbide coating on the four pack coated, four out-of-pack coated and four gas CVD coated samples to be used in the abrasive wear tests was determined by ball-cratering. Further, the hardness of the four uncoated samples to be used in the tests and of the substrate of each of the coated samples was determined by Vickers hardness tests, subsequently performed on the area of substrate exposed by ball-cratering in the case of the coated samples. Finally, the thickness of each sample itself was measured. The values obtained are detailed in Table 1. It is evident from Table 1 that comparable coating thicknesses have not been achieved in the three coating processes. As will be described, this did not, however, preclude the valid use of the coated samples in the abrasive wear tests.

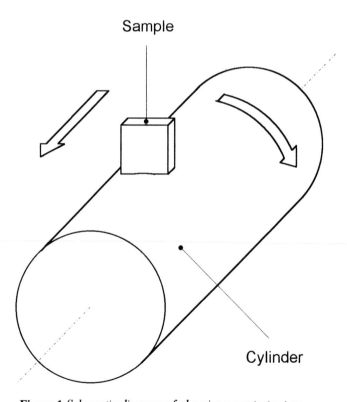

Figure 1 *Schematic diagram of abrasive wear test set-up*

2.2 Abrasive Wear Tester

A modified Myford "Super 7" lathe was employed to carry out the abrasive wear tests. The sample to be tested was held in a pivoted arm mounted on the lathe cross-slide. A cylinder, consisting of a mild steel bar spirally wrapped with P1200 grade silicon carbide abrasive paper (total diameter 51.25 mm), was held between centres on the lathe with a driving dog attached to it at the headstock end. The pivoted arm could be located in two positions. In the test position, a side face on the sample to be tested was held against the cylinder, the long edges of this face being aligned perpendicular to the cylinder axis. During testing, the cylinder was rotated in the lathe headstock and an axial feed engaged so that the sample traced a spiral path along the cylinder. The set up is illustrated schematically in Figure 1. The placement of a dead weight on the pivoted arm, together with counterbalancing of the arm to varying degrees, facilitated the application of a wide range of loads. Rotating the pivoted arm through 180° located it in the measurement position, in which a low-power microscope could be used to measure the width of the wear scar on a sample without disturbing its position.

2.3 Abrasive Wear Tests

The standard conditions used in the abrasive wear tests were as follows: applied load 44 g, cylinder rotational speed 25 rev min^{-1} and feed applied to the sample 1.52 mm rev^{-1}. These standard conditions were established through preliminary experimentation. In the case of the chromium carbide coated samples, it was anticipated that an increase in abrasive wear rate would occur when the coating was worn through, i.e. when the depth of the wear scar exceeded coating thickness. Consequently, further preliminary experimentation was carried out to establish the approximate sliding distance corresponding to this point for one each of the pack, out-of-pack and gas CVD coated samples. Not surprisingly, in view of their significantly

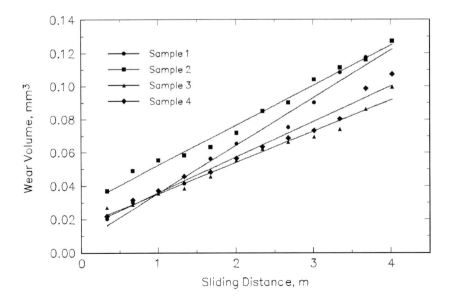

Figure 2 *Variation of wear volume with sliding distance for uncoated samples*

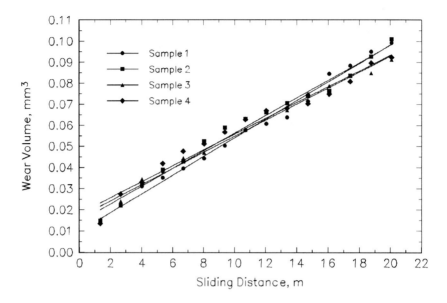

Figure 3 *Variation of wear volume with sliding distance for pack coated samples*

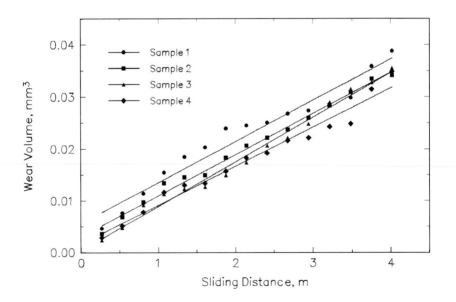

Figure 4 *Variation of wear volume with sliding distance for out-of-pack coated samples*

different coating thicknesses (Table 1), the sliding distance corresponding to coating break-through was found to differ substantially for each type of coated sample, being longer the thicker the coating as would be expected. The wear tests with each type of coated sample were subsequently carried out for a sliding distance of approximately twice that corresponding to coating breakthrough in each case. In these tests and those with the uncoated samples, testing was interrupted at regular intervals and the width of the wear scar measured using the aforementioned low-power microscope. The measurement intervals were chosen so as to ensure an adequate number of data points, both before and after coating breakthrough in the case of the coated samples. Each piece of abrasive paper used in the tests was discarded after one traverse along its length.

On completion of the tests, the abrasive wear volume pertaining to each wear scar width measurement was calculated. Figures 2, 3, 4 and 5 respectively show the variation of wear volume with sliding distance thus determined for the uncoated, pack coated, out-of-pack coated and gas CVD coated samples. As is clearly evident from each of the latter three of these Figures, the anticipated increase in abrasive wear rate when the depth of the wear scar exceeded coating thickness (i.e. at approximately half the total sliding distance in each case) did not materialise for any of the coated samples. This being the case, linear regression analysis was employed to fit the best straight line to the whole of each of the wear curves, with the exception of the small, initial period of breaking-in wear. From this analysis, the slope of the line of best-fit, i.e. the steady state abrasive wear rate, and the correlation coefficient were calculated for each wear curve, the values obtained being shown in Table 2. The grouped bar chart in Figure 6 shows the different sample types ranked in, apparently, decreasing steady state abrasive wear rate order.

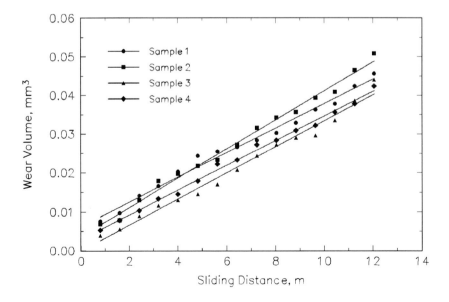

Figure 5 *Variation of wear volume with sliding distance for gas CVD coated samples*

Table 2 *Steady state abrasive wear rates and correlation coefficients for uncoated and chromium carbide coated samples*

Sample number	Steady state wear rate mm^3 mm^{-1}	Correlation coefficient
	Uncoated samples	
Sample 1	2.833×10^{-5}	0.988
Sample 2	2.413×10^{-5}	0.994
Sample 3	1.888×10^{-5}	0.988
Sample 4	2.144×10^{-5}	0.989
	Pack coated samples	
Sample 1	4.426×10^{-6}	0.996
Sample 2	4.186×10^{-6}	0.991
Sample 3	3.818×10^{-6}	0.992
Sample 4	3.757×10^{-6}	0.985
	Out-of-pack coated samples	
Sample 1	7.923×10^{-6}	0.975
Sample 2	7.931×10^{-6}	0.996
Sample 3	8.597×10^{-6}	0.994
Sample 4	7.535×10^{-6}	0.986
	Gas CVD coated samples	
Sample 1	3.165×10^{-6}	0.993
Sample 2	3.757×10^{-6}	0.996
Sample 3	3.354×10^{-6}	0.992
Sample 4	3.182×10^{-6}	0.997

3 DISCUSSION OF RESULTS

As described earlier, the anticipated increase in abrasive wear rate of the chromium carbide coated samples when the depth of the wear scar exceeded their coating thickness, did not materialise. Consequently, linear regression analysis was employed to fit the best straight line to the whole of each of the wear curves for the coated samples (except for the small, initial period of breaking-in wear). In initial justification of this action, it can be seen from Table 2 that correlation coefficients > 0.99 were thus obtained for 75% of the wear curves for the coated samples, whereas the corresponding figure for the uncoated samples, where, of course, the concept of an increase in wear rate due to coating break-through has no relevance, was only 25%. It is tentatively suggested that this unexpected abrasive wear test response exhibited by the coated samples is due, as illustrated schematically in Figure 7, to the worn coating remaining at the entrance to and exit from the abrasive wear scar, effectively acting as "bearing surfaces" for the cylinder, bridging the gap in the coating and thereby maintaining the wear rate at a level characteristic of the coating. In defence of this suggestion, a somewhat similar phenomenon is observed in the case of hard material coated cemented carbide cutting tools. It is widely found that although the crater wear rate of such tools increases when the coating is worn through on their rake face, it does not increase to a level characteristic of the cemented carbide substrate. An extensively quoted explanation for this is that after coating breakthrough the chip is supported by the worn coating remaining at the periphery of the crater[9,10] The difference in the effect on wear rate after breakthrough of the chromium carbide coatings in

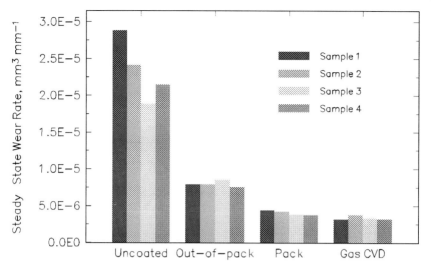

Figure 6 *Steady state abrasive wear rates of uncoated, out-of-pack coated, pack coated and gas CVD coated samples*

the abrasive wear tests of the present work and of the hard material coatings on cemented carbide cutting tools may be associated with the significantly higher contact temperatures, stresses and strain rates pertaining in the case of the latter.

Abrasive wear can take place either by plastic deformation or by brittle fracture. They often occur in concert[11]. The latter is, however, known to be suppressed by low loads or small abrasive particles. As detailed earlier, both are germane to the current abrasive wear tests. This, together with the appearance of the worn surfaces of the uncoated and coated test

Figure 7 *Schematic diagram showing coated sample after coating breakthrough*

samples, suggests that abrasive wear by plastic deformation, as opposed to brittle fracture, is likely to have predominated in the present work. The simple theory of abrasive wear by plastic deformation predicts, as can be seen from the equation for it below[11,12] (in which V is the wear volume, S is the sliding distance, W is the applied normal load, H is the hardness of the softer material and k is a dimensionless wear coefficient), that abrasive wear volume should increase linearly with sliding distance. In practice, however, this is found to be the case only if fresh abrasive particles are continually being presented[12].

$$\frac{V}{S} = k\frac{W}{H} \qquad\qquad \{1\}$$

Since this criterion was (deliberately) met in the present work, by virtue of the spiral path traced by a sample along the cylinder, then the linear increase in wear volume with sliding distance, following the small, initial period of breaking-in wear, for both the uncoated and chromium carbide coated samples manifest in Figures 2 to 5 is entirely consistent. Substituting the normal load applied and the relevant values of uncoated sample hardness, hardness of the pack, out-of-pack and gas CVD chromium carbide coatings (1520, 1460 and 1570 $HV_{0.025}$ respectively[8]) and steady state wear rate into equation {1}, dimensionless wear coeffcient values can be calculated for the present work. These calculated values are presented in Table 3. Typical dimensionless wear coefficients for two-body abrasive wear by plastic deformation[12] are between 2×10^{-1} and 2×10^{-2}. The k values calculated in the present work for the pack and gas CVD coated samples can be seen from Table 3 to lie wholly within this range, whilst those for the out-of-pack coated and, particularly, the uncoated samples lie outside but close to its upper end. This is, therefore, seen as adding weight to the suggestion that abrasive wear by plastic deformation, as opposed to brittle fracture, is likely to have predominated in the present work. It is also viewed as further justifying the action of fitting the best straight line to the whole of the wear curve for each of the coated samples (except for the small, initial period of breaking-in wear).

From Table 2 and Figure 6, it appears to be the case that the abrasive wear rate of the samples studied in the present work decreases in the order: uncoated, out-of-pack coated, pack coated and gas CVD coated. However, as only four samples of each type were tested, it was thought advisable to analyse the results statistically in order to obtain the most meaningful interpretation of them. The t-test[13] was thus employed to compare mean wear rates. Using this test it was established at a 99.9% confidence level that the mean wear rate of the pack coated samples was different to that of the out-of-pack coated samples, and at a 99% confidence level that the mean wear rate of the out-of-pack coated samples was different to that of the uncoated samples. At a 99% level of confidence, however, no significant difference in the mean wear rates of the pack and gas CVD coated samples was established. In summary,

Table 3 *Dimensionless wear coefficient values for uncoated and coated samples*

Sample type	Dimensionless wear coefficient, k
Uncoated	$2.74 \times 10^{-1} - 4.19 \times 10^{-1}$
Out-of-pack coated	$2.50 \times 10^{-1} - 2.85 \times 10^{-1}$
Pack coated	$1.30 \times 10^{-1} - 1.53 \times 10^{-1}$
Gas CVD coated	$1.14 \times 10^{-1} - 1.34 \times 10^{-1}$

therefore, it is concluded that the mean wear rate of the pack coated samples is less than that of the out-of-pack coated samples, which, in turn, is less than that of the uncoated samples, but that there is no significant difference between the mean wear rates of the pack and gas CVD coated samples.

It is evident from equation {1} that, according to the simple theory of abrasive wear by plastic deformation, the wear volume should decrease linearly with increase in the hardness of the softer material. Since all three types of chromium carbide coatings had a hardness (quoted earlier) approximately 2.5 times greater than that of the uncoated samples (see Table 1), then it might simplistically have been anticipated that the abrasive wear rate of the chromium carbide coated samples would have been of the order of 2.5 times less than that of the uncoated samples. In reality, however, it can be seen from Table 2 that the wear rate of the out-of-pack coated samples is approximately 3 times less than that of the uncoated samples, whilst the wear rate of the pack coated and gas CVD coated samples is approximately 6 times less than that of the uncoated samples. The reason for this apparent discrepancy is thought to lie chiefly in the fact that whilst there are, generally, inverse correlations between abrasive wear rate and hardness within generic groups of materials, the same correlation is not found in different groups[14]. Since, self evidently, AISI 44OC martensitic stainless steel and chromium carbide do not belong to the same generic material group, then it is not realistic to infer that their relative wear rates, under conditions of two-body abrasive wear by plastic deformation, will be equal to the inverse ratio of their hardnesses.

That there does not appear to be a significant difference between the reduction in abrasive wear rate caused by the pack and gas CVD chromium carbide coatings is consistent with the fact that, with three relatively small exceptions (bulk grain structure, surface appearance and

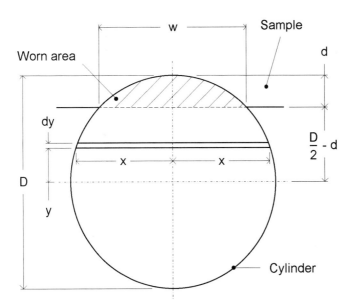

Figure 8 *Schematic diagram showing worn area on a sample*

scratch adhesion test behaviour), the characteristics of these two coatings were established, in the detailed study preceding the present work[8], to be basically equivalent. In contrast, several of the out-of-pack chromium carbide coating characteristics were found to not only differ from, but also to be inferior to those of the pack and gas CVD coatings. In particular, the structure of the out-of-pack coating, both at its surface and for some considerable distance into its bulk, was observed to be less dense than that of the pack and gas CVD coatings. It is thought likely that this is largely responsible for the significantly smaller reduction in abrasive wear rate brought about by the out-of-pack chromium carbide coating, in comparison to the pack and gas CVD coatings.

4 CONCLUSIONS

The major conclusions arising from the present work are detailed below. It is not yet known if they are specific to the abrasive wear test conditions employed in the present work or if they have wider relevance.

1. At a 99% level of confidence, it can be stated that the mean abrasive wear rate of the pack chromium carbide coated AISI 44OC martensitic stainless steel samples is less than that of the out-of-pack coated samples, which, in turn, is less than that of the uncoated samples, but that there is no significant difference between the mean wear rates of the pack and gas CVD coated samples.
2. The wear rate of the out-of-pack coated samples is approximately 3 times less than that of the uncoated samples, whilst the wear rate of the pack and gas CVD coated samples is approximately 6 times less than that of the uncoated samples.
3. The fact that there does not appear to be a significant difference in the reduction in abrasive wear rate caused by the pack and gas CVD coatings is consistent with the characteristics of these two coatings, with three relatively small exceptions, having previously been established to be basically equivalent. The significantly smaller reduction in abrasive wear rate brought about by the out-of-pack coating can, it is thought, be largely attributed to the structure of this coating, both at its surface and for some considerable distance into its bulk, being less dense than that of the pack and gas CVD coatings.
4. The results of the present work suggest that the growing trend for the pack cementation process to be supplanted by (conventional, open-reactor) gas CVD for the deposition of chromium carbide coatings, does not have prejudicial consequences with regard to abrasive wear resistance. At the present stage of its development, however, the use of the out-of-pack cementation process for chromium carbide coating deposition does appear to be inimical in this respect.

Acknowledgements

The authors wish to thank Loughborough University of Technology for the provision of research facilities and the Directors of Diffusion Alloys Limited for their permission to publish this paper.

References

1. E. Horvath and A.J. Perry, *Wear*, 1978, **48**, 217.
2. B. Eastwood, S. Harmer, J. Smith and A. Kempster, *Materials Science Forum*, 1992, **102–104**, 543.
3. W. Hanni and H. E. Hintermann, *Thin Solid Films*, 1977, **40**, 107.
4. K. Jyrkas and L. H. J. Lajunen, *Surface Engineering*, 1990, **6**, 113.
5. A. J. Perry and E. Horvath, *J. Materials Science*, 1978, **13**, 1303.
6. R. L. Samuel and N. A. Lockington, Metal Treatment and Drop Forging, 1952, 19581.
7. H. E. Hintermann, A.J. Perry and E. Horvath, *Wear*, 1978, **471**, 407.
8. A. B. Smith, A. Kempster, J. Smith and G.W. Critchlow, Proceedings of the 13th International Plansee Seminar, eds. H. Bildstein and R. Eck, Metallwerk Plansee, 1993, Volume 3, p. 129.
9. B. N. Colding, S.M.E. Technical Paper, MR80–901, 1980, 1.
10. D. E. Graham and T. E. Hale, *Carbide and Tool Journal*, 1982, 34.
11. I. M. Hutchings, 'Tribology – Friction and Wear of Engineering Materials', Edward Arnold, 1992.
12. E. Rabinowicz, 'Friction and Wear of Materials', John Wiley & Sons, 1965.

1.3.4

Abrasion Wear Behaviour of Squeeze Cast SiC Particulates Reinforced Al-2024 Composites

S. M. Skolianos, G. Kiourtsidis and L. Escourru

LABORATORY OF PHYSICAL METALLURGY, DEPARTMENT OF MECHANICAL ENGINEERING, ARISTOTLE UNIVERSITY OF THESSALONIKI, GREECE

1 INTRODUCTION

Wear is one of the main causes which lead to a rather high number of annual losses in the industrialized nations. As a result there is an increasing need for materials with very good wear properties. A candidate type of material is metal matrix composites. SiC-reinforced aluminum alloy matrix composites are of especial interest since for a number of applications – ranging from pistons and cylinder liners to lightweight bearing sleeves and brake rotors – the tribological properties of this kind of composites are important[1-3].

One way to produce metal matrix composites is through solidification processing which is a rather simple and economical technique. The main advantage of the casting process, especially stir casting, is high productivity and reduced cost. In stir casting the ceramic particulates are added to a partially solidified matrix alloy then the semi-solid slurry is vigorously agitated before casting in moulds[4-5]. In this technique the high effective viscosity of the slurry prevents the ceramic particles from floating, settling or agglomerating and by increasing the mixing time after addition, interaction between the particles and the liquid alloy matrix promotes bonding. If this technique is combined with squeeze casting near-net castings can be produced. The solidification rate affects the microsegregation and coarseness of the dendrite matrix, the amount of microporosity, the distribution of the ceramic particles and ultimately the mechanical behaviour of the cast material.

The objective of this work was the determination of the abrasive wear behaviour of squeeze cast SiC particulates reinforced Al-2024 composites.

2 EXPERIMENTAL PROCEDURE

2.1 Fabrication of the Composites

The specimens for abrasive testing were fabricated by the stir casting technique. The process was basically similar to that described earlier[5-7]. The alloy was melted at about 800°C and then the melt was slowly cooled to 632°C. This temperature lies in the solid/liquid range and corresponds to a volume fraction of the primary α-phase of about 0.42. During the cooling the metal was vigorously stirred at about 1000 rpm with a carbide impeller, which could be lifted or lowered inside the crucible and was driven by a variable motor. In this way a semi-solid metal slurry was created. After a five minute isothermal holding the SiC particles, which were preheated at 400°C for 12 h, were gradually added in the vortex created by the stirring and the stirring of the mix was maintained for about 30 minutes. The temperature was held at 633–640°C i.e. just below the liquidus temperature of the alloy which was found experimentally to be 650°C. Then the melt was reheated to above the liquidus temperature, at 750°C, and then squeeze cast, under a pressure of 100 MPa, into a preheated (400°C) cylindrical mould. During the melting Borax powder was added to the melt following standard foundry practice in order to clean the melt. The impeller was completely immersed in the slurry and its axis of rotation was parallel to the crucible axis.

2.2 Abrasion Testing

Abrasion tests were performed using a pin-on-disc type apparatus. All the tests were performed in air and at room temperature. The samples were cylinders cut from the ingots with a diameter of about 0.007 m. With the pin lowered a static weight was added to the holder. The weights ranged from 0.5 to 2.9 kg. A spiral spring was used to smooth the procedure by neutralizing the dynamic oscillations which appear during the experiments. The abrasive counterface used, was P60 grade grit silicon carbide commercial paper. Three tangential velocities were used, 0.18, 0.26 and 0.33 ms^{-1}. Each test was periodically interrupted to weigh the sample and avoid heating. The specific wear rate, W_s (mm^{3}N^{-1}m^{-1}), and the weight loss (gr) were used to report the wear data, where, $W_s = \Delta_m/\rho AP_nL$, where Δ_m is the mass loss [g], ρ the density of the material [g/mm^3], A the contact surface [mm^2], L the sliding distance [m] and P_n the plane stress equal to F_n/A where F_n is the load vertical to the contact surface.

3 RESULTS AND DISCUSSION

3.1 Structures of the Composites

The stir cast microstructure of the composite is shown in Figure 1, for the squeeze cast and the conventional cast alloy. The light areas denote the primary α-phase particles of the matrix alloy. The ceramic particles are rather homogeneously distributed in the alloy. A binary eutectic α/θ–Al$_2$Cu occupies the interdendritic spaces. In some areas some small amounts of ternary eutectic α/θ/Al$_2$CuMg can be seen. Particles of Mg$_2$Si are occasionally observed around the reinforcing particles, as the result of the reaction of the magnesium with SiO$_2$ which is always present on the carbide surface.

During processing of these composites reaction between silicon carbide particulates and aluminum yields Al$_4$C$_3$, leading to degradation of the reinforcing phase. This reaction is particularly encouraged by the absence of silicon from the alloy and was described previously[8,9]. At higher cooling rates the silicon carbide particles are partially trapped by the moving interface while at lower cooling rates they appear to be pushed by the advancing solid/liquid interface and to gather in the interdendritic spaces, where coarse intermetallic particles appear. This affects the fracture behaviour of the material[8, 9].

The principal difference in the microstructure of the squeeze cast and the conventional cast alloy is porosity, Figures 1 and 2. These micrographs stress the beneficial effect on microstructure of squeeze casting. In the squeeze cast alloy no pores are present in contrast to the conventionally cast alloy which exhibits porosity to a rather large extent. It is clear from

(a)

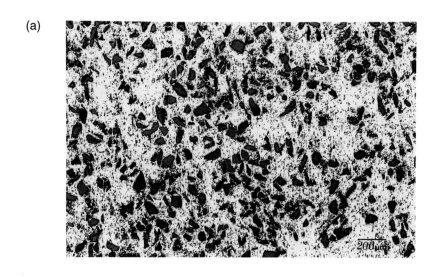

(b)

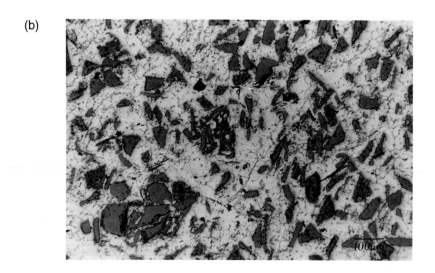

Figure 1 *Microstructure of squeeze cast SiC$_p$-reinforced Al-2024 alloy specimen. Volume fraction (a) 0.07, (b) 0.17*

(a)

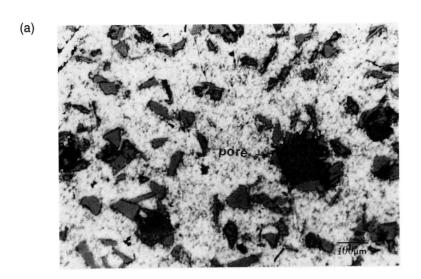

(b)

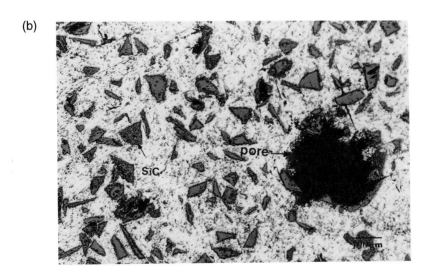

Figure 2 *Microstructure of conventional cast SiC$_p$-reinforced Al-2024 alloy specimen.
Volume fraction (a) 0.07, (b) 0.17*

these micrographs that porosity increases and pores become larger as the volume fraction of the SiC particulates increases. This can be attributed to the fact that apart from normal casting porosity resulting from dissolved gases or shrinkage, there may be additional porosity due to the stir casting process. During agitation of the semi-molten metal slurry an intimate interaction is developed between the molten alloy and environmental gases which enhances the dissolution of gases. Furthermore the formation of the vortex may lead to suction of air bubbles together with the SiC powder[10]. As a result bubbles-particles are combined and porosity nucleates heterogeneously on the surface of the dispersed ceramic particles. These mechanisms are mainly responsible for the observed porosity at the particle-matrix interface, Figures 1 and 2.

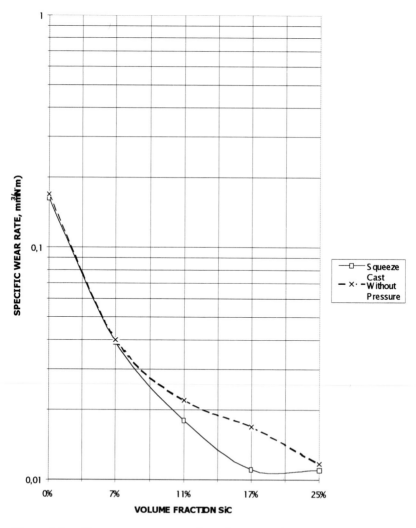

Figure 3 *Specific wear rate versus volume fraction of SiC$_p$ dispersed in Al-2024 alloy matrix. Average SiC particle size 40 μm*

Thus, the extent of porosity depends on the amount of the reinforced particles, as long as the other process variables, such as environmental conditions, stirrer speed, size and position, remain constant. This is illustrated in Figures 1 and 2, where it is apparent that in the conventional alloy the size of pores is not very large and porosity is kept in rather low values inasmuch as the volume fraction of SiC remains rather low. But as the amount of SiC increases porosity increases and pores become quite large. However, if squeeze cast is applied no porosity is observed, Figure 2.

3.2 Abrasive Wear

Measurements of abrasive wear resistance of various specimens by the pin-on-disc tribometer are plotted in Figure 3 as specific wear rate W_s (mm³N⁻¹m⁻¹) vs. volume fraction carbide measured metallographically. Figure 3 compares the behaviour of particulate composites with an Al-2024 matrix which were squeeze cast and those which were conventionally cast. For both types of composites the reinforcing carbide has an average particle size of 40 μm. The specific wear rate decreases with increasing carbide volume fraction and reaches a minimum at 17% for the squeeze cast alloy. Beyond that point it appears to remain constant. In the absence of reinforcement, the specific wear rate of the Al-2024 alloy is about 1.6×10^{-1} mm³N⁻¹m⁻¹, whereas that of the squeeze cast composite with a volume fraction SiC_p 17% is 1.1×10^{-2} mm³N⁻¹m⁻¹. Squeeze casting does not affect significantly the wear of the unreinforced alloy and of the composites up to a volume fraction of 7%. At such a low volume fraction of reinforcing phase porosity is rather small in the conventional alloy, Figure 1, and it does not appear to affect the wear behaviour. However as the SiC_p volume fraction increases between 7% and 25%, porosity increases considerably in the conventional alloys, Figure 2, and so squeeze cast specimens exhibit a lower abrasive wear rate than that of the conventionally cast alloys. The difference between them appears to vanish at the SiC_p volume fraction of 25%. At such a high SiC_p volume fraction the interparticular matrix would be so thin that the mechanism

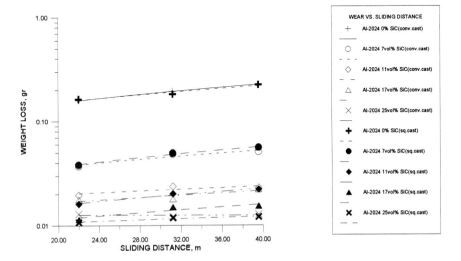

Figure 4 *Weight loss versus sliding distance of SiC_p-reinforced Al-2024 alloy specimens. Average SiC particle size 40 μm*

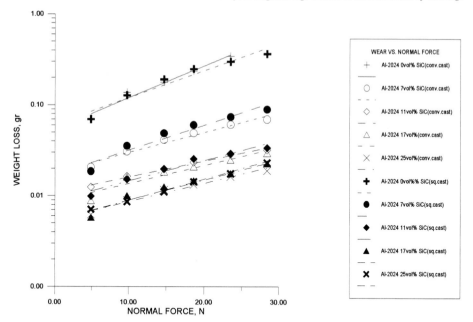

Figure 5 *Weight loss versus applied normal force of SiC$_p$-reinforced Al-2024 alloy specimens. Average SiC particle size 40 μm*

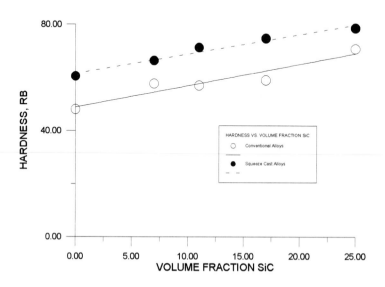

Figure 6 *Hardness versus volume fraction of SiC$_p$ dispersed in Al-2024 alloy matrix. Average SiC particle size 40 μm*

of abrasive wear in this composite appears to be chipping of the carbide particles instead of plastic ploughing in the interparticle matrix which was the main mechanism at lower SiC_p volume fractions. Therefore porosity does not influence the wear behaviour.

The wear behaviour of the Al-2024 alloy and of the composites as weight loss (gr) versus sliding distance is shown in Figure 4, for both the squeeze cast alloys and the conventionally cast alloys. Apart from the composite with a SiC_p volume fraction of 25%, which exhibits an almost constant wear, all other specimens display an increase in weight loss with the sliding distance. The squeeze casting does not affect the wear of the unreinforced alloy and of the composites with 7 and 25 vol.% SiC_p. However there is a significant difference for the 17 vol.% composite. In this case squeeze casting improves wear behaviour. With a volume fraction of 11% SiC_p the squeeze cast alloy exhibits slightly better wear behaviour than the conventional one for low values of the sliding distance. As the sliding distance increases the effect of the squeeze casting diminishes. These results are consistent with the above analysis.

The effect of the applied load on the wear behaviour is illustrated in Figure 5 where the variation of weight loss with the applied normal force has been compared for unreinforced, reinforced, squeeze and conventionally cast alloys. With increasing load the weight loss increases in all the alloys. The effect of squeeze casting is similar to that described before for the relationship between the wear and the sliding distance. Again a significant improvement is observed in the squeeze cast reinforced Al-2024 composite with 17 vol.% SiC_p while in all the other cases the results are comparable.

Hardness may be the mechanical property that can best be correlated with abrasive wear resistance. The dependence of Rockwell hardness on SiC_p volume fraction for squeeze and conventional cast composites is represented in Figure 6. For each specimen several measurements were taken randomly across the specimen surface and averaged yielding the hardness of the composite. As expected, the hardness increases with volume fraction of the harder phase. It is also higher in the squeeze cast specimen.

4 CONCLUSIONS

1. The specific wear rate decreases and the composite hardness increases with increasing SiC_p volume fraction.
2. Squeeze casting leads to an increase in wear resistance as long as porosity is substantial in the conventional alloys.
3. Squeeze casting improves the hardness of the composites.

Acknowledgments

The work was supported by a grant from the General Secretariat of Research and Technology, Ministry of Industry of Greece.

References

1. T. I. Ho and M. B. Peterson, *Wear*, 1977, **43**, 199.
2. S. V. Prasad and P. K. Rohatgi, *J. of Metals*, 1987, **11**, 22.

3. P. K. Rohatgi, *J. of Metals*, 1991, **43**, 10.
4. R. Mehrabian, R. Riek and M.C. Flemings, *Met. Trans.*, 1974, **5**, 1899.
5. C. G. Levi, G. S. Abbaschian and R. Mehrabian, *Met. Trans. A*, 1978, **9A**, 697.
6. A. Sato and R. Mehrabian, *Met. Trans. B*, 1976, **7B**, 443.
7. S. Skolianos and T. Z. Kattamis, *Mater. Sci. Eng.*, 1993, **A163**, 107.
8. D. J. Lloyd, *Composites Science and Technology*, 1989, **35**, 159.
9. D. J. Lloyd, H. Lagace, A. McLeod and P. L. Morris, *Mater. Sci. Eng.*, 1989, **A107**, 73.
10. S. Ray, in 'ASM Proceedings on Cast Reinforced Metal Composites', edited by S.G. Fishman and A.K. Dhingra (ASM, Metals Park, Ohio, 1988) p. 77.

1.3.5

Influence of Thickness of Electroless Nickel Interlayer on the Tribological Behaviour of Mild Steel Coated With Titanium Nitride

C. Subramanian and K.N. Strafford

IAN WARK RESEARCH INSTITUTE, UNIVERSITY OF SOUTH AUSTRALIA, THE LEVELS, S.A. 5095, AUSTRALIA

1 INTRODUCTION

The life of engineering components and tools is often influenced by surfaces which can now be tailored to meet specific requirements in terms of wear and corrosion resistance, and to some extent, fatigue. Another area where surfaces play an important rôle is in the bio compatibility of surfaces of artificial materials inserted in biological systems, e.g. total hip joint replacement. Recent years have witnessed an explosion of published information in this area – surface engineering – which includes both traditional technologies such as heat treatment and new ones such as plasma-based physical and chemical vapour deposition and ion implantation. Titanium nitride coatings are commercially very successful in applications involving cutting tools, metal forming dies and tools, plastic moulding tools, etc. Other non-engineering applications are in decorative coatings and jewellery due mainly to its golden colour. The overwhelming lead of TiN over other hard coatings can be attributed to its colour even in applications where colour is not a selection criterion.

TiN coatings can be produced by either physical vapour deposition (PVD) or chemical vapour deposition (CVD) for different applications. The substrates on to which these coatings are deposited vary depending on the application. For example, the cutting tool industry generally uses a CVD method to apply TiN coatings on carbide tools whereas the metal forming industry coats die/tool steel substrates with TiN by PVD. There have been few attempts to deposit TiN on lower grade steel substrates such as low alloy steel or mild steels. The main drawback in such coating systems is the load-bearing capacity of the substrate. When TiN coatings are deposited on softer substrates the coatings collapse due to what is termed "eggshell effect". This leads to poor adhesion of coatings to the substrate resulting in spallation of coatings even under mild wear conditions.

To counter the spallation of coatings due to the softness of the substrate, the latter should be strengthened by heat treatment such as nitriding and/or an interlayer deposited prior to TiN coating. The interlayer should be compatible with not only the substrate but also the coating. Previous reports discuss the use of interlayers such as Cr[1], Ti[2,3] and Ni[4-6] with and without nitriding of the substrate.

Subramanian et al[1] investigated the adhesion and sliding wear behaviour of TiN-coated low alloy steel (AISI 4150) with chromium as an interlayer. The substrate was hardened by nitriding prior to Cr plating and TiN coating. It was found that the TiN plus Cr plating on a nitrided steel was most effective in promoting adhesion and wear resistance.

Ti interlayers between steel substrates and TiN coatings have been employed by many workers in promoting adhesion[2,3]. Ti is an obvious choice as an interlayer material because TiN coating can be deposited after Ti in the same process cycle in the same chamber. However, the lack of hardness in the Ti layer resulted in poor overall load-bearing capacity. This has led to the use of electroless nickel which is harder than Ti interlayer.

Electroless Ni, unlike pure Ni by electrolytic plating, is an alloy of Ni and P[7,8]. The pH of the hypophosphite-based bath controls the level of P in the coating and thus its microstructure. The structure of the electroless nickel coatings changes from amorphous to crystalline as the P content increases in the coating. The coating can be hardened by heat treatment by the precipitation of Ni_xP_y particles.

Electroless nickel interlayers have been applied to mild steel and tool steel substrates prior to TiN coating[4-6]. Chen and Duh[4] have shown that the electroless interlayer crystallised during sputtering and the hardened interlayer improved the adhesion and overall hardness of the coating.

He and Hon[5] have shown that by placing an electroless nickel interlayer, it is possible to improve the adhesion and wear resistance of TiN coated tool steel. Doong and Duh[6] have studied the influence of pH of the plating bath (and thus the levels of P) on the adhesion and hardness of electroless nickel interlayers in a TiN coated mild steel system. Although the range of thickness investigated by Doong and Duh[6] was limited to 0–12μm, it was possible to show the influence on hardness of the total coating system of thickness of electroless nickel interlayer within the range studied.

This paper discusses the influence of the thickness of the Ni–P alloy interlayer on the hardness and adhesion of the TiN coatings on mild steel substrates. The range of thickness of electroless nickel interlayers used was 0–104.3 μm. The mild steel substrates were used in two different heat treated conditions namely, quenched, and quenched and tempered.

2 EXPERIMENTAL

AISI 1030 steel was chosen as the substrate material which was used in two conditions – as-received and quenched. The quenched specimens (75 mm diameter x 5 mm thick) were heat treated at 850°C for an hour and quenched in water. The substrates were coated with electroless nickel (EN) with varying thickness (0 - 104.3μm). The actual (measured) thickness of the EN coatings were 2.1, 2.7, 5.3, 15.7, 23.0, 49.0, 104.3 mm. The pH of the electroless nickel bath was adjusted such that the EN coatings had a P content of 6 wt. %. TiN was deposited using a commercially available PVD unit (Surface Technology Coatings, Melbourne, Australia).

A VTT Scratch Tester (VTT Technology Inc., Espoo, Finland) was used to evaluate the adhesion of TiN coatings. The critical load (Lc) for the removal of the coating was used as an indication of coating adhesion. The test involved traversing a diamond stylus with a 0.2 mm tip radius across the TiN coated surface under a continuously increasing normal load. During the scratch tests, the acoustic emission and tangential force (frictional force) values were recorded. Acoustic emission, change in the slope of the frictional force (i.e. change in coefficient of friction) and microscopic examination were all employed in the accurate determination of the actual values of Lc. It appeared that the change in the friction coefficient as a function of sliding time (and applied load as the load was continuously varied) was a better indicator. The scratch tests were carried out at a speed of 10 mm/min and at a loading rate of 50 N/min. The maximum load applied was restricted to around 80 N.

A ball-on-disc sliding wear tester was employed to evaluate the wear resistance of these EN-TiN coatings. The counter-surface used was yttria-stabilised tetragonal zirconia (YTZ) balls of 10 mm diameter, supplied by Tosoh Company of Japan. The balls and discs were thoroughly cleaned, degreased and dried before weighing. The accuracy of weighing was ± 0.01 mg. The disc was rotated in contact with the ball under a load of 50 N at a sliding speed of 0.3 m/s in ambient atmosphere (~20°C). The test duration was generally 1 h. After the test, both the disc and the ball were weighed to determine the weight loss, followed by scanning electron microscopy (SEM) of the wear tracks.

3 RESULTS AND DISCUSSION

3.1 Hardness

The Vickers hardness values for EN and EN+TiN coatings on mild steel in the as-received and quenched conditions are shown as a function of the thickness of the EN interlayer in Figure 1. The hardness of EN coatings on mild steel substrate increases with the thickness of EN layers from 245 to 550 HV. When both EN and TiN are present, the "composite" hardness of the double layer also increases from 455 to 1300 HV on the as-received mild steel and from 550 to 1230 HV on the quenched mild steel. It should be noted that when thicker EN layer is present the substrate microstructure/hardness becomes less critical. Further, when the substrate is quenched (i.e. harder), the composite hardness reaches its peak value when the EN layer is less than 10 μm whereas for as-received mild steel a thicker EN layer is required. In other words, the critical thickness of the EN layer depends on the hardness of the substrate.

3.2 Adhesion

The adhesion of TiN coatings to its substrates is generally measured by scratch testing. The critical load to failure depends on many factors such as the hardness of the substrate, the thickness of TiN coatings and surface finish[9]. Keeping other factors constant, it is possible to compare the adhesion of a number of TiN coatings. This is useful in a production situation where the scratch testing can be used as a quality control tool. It should also be mentioned that the critical load to failure is not only an indicator of adhesion but also parameter related to the load-bearing capacity of the complete coatings system. This information would be very useful in tribological applications.

Figure 2 shows the variation of the critical load with the thickness of the EN interlayer for the substrate in two conditions – as-received and quenched. For both the cases, Lc increases with the thickness of EN layer. The rate of increase was higher for the thin than the thick EN layers. The critical thickness for the cases appear to be around 10 μm. Again, the harder substrate does not require the same thickness of interlayer as the soft substrate.

3.3 Friction

The coefficient of friction was calculated from the tangential force measurements recorded during the scratch tests. The tangential force traces had two distinct regions – below and above Lc. Figure 3 shows the variation of friction coefficient values for as-received and quenched mild steel substrates coated with EN and TiN at loads below and above Lc. Friction

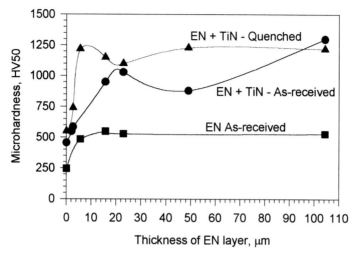

Figure 1 *Microhardness of electroless nickel (EN) and EN+TiN coatings on mild steel substrate*

coefficient values decrease with the increase in the thickness of EN interlayer for all cases – below and above Lc and for the as-received and quenched substrates. For applied loads below Lc, there is not a significant difference between the friction coefficient values of EN+TiN coatings on as-received and quenched substrates. On the other hand, for applied loads above Lc, the coatings on as-received substrates show higher coefficient of friction than for coatings on quenched substrates. This is probably due to the mechanism of wear of the coatings as the load was increased. When the load in increased, obviously the top layer of TiN is penetrated.

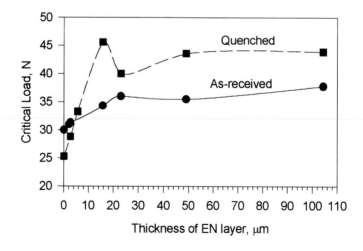

Figure 2 *The critical load to failure in scratch testing as a function of the thickness of EN layer*

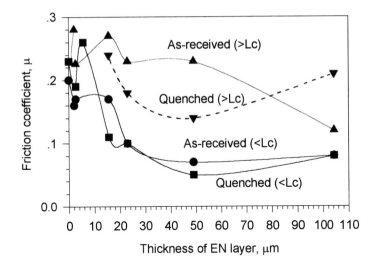

Figure 3 *Friction coefficient values for as-received and quenched mild steel substrates, coated with EN+TiN coatings at loads below and above Lc*

Once the TiN is penetrated the friction increases as the surface on which the diamond stylus slides has changed from TiN to EN layer, with a minor contribution from TiN.

3.4 Wear

The wear rates of the EN-TiN coatings as a function of EN layer thickness are shown in Figure 4. It appears that the wear rate increases with the thickness of EN layer up to 23 μm. The lowest wear rate was observed for TiN coating with 104.3 μm EN layer. The question

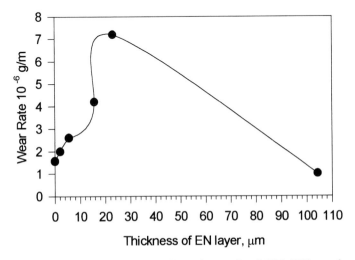

Figure 4 *The wear rate of quenched mild steel coated with EN+TiN as a function of thickness of EN layer (5 kg load; 0.3 m/s)*

here is "Why does the wear rate increase with the thickness of EN layer when the adhesion (and to some extent, the load-bearing capacity) of the EN–TiN coatings increase?" Wear is a very complex phenomenon and it is influenced by not only material properties but also the tribological (testing) system parameters. Further research is needed with these coatings under varying loads to confirm any change in the critical load or thickness at which the wear mechanism changes from one to another. It is possible that there is a transition in the load-bearing capacity, adhesion and friction around 15–25μm of the EN layer. Further research on these duplex coating systems is focussed on the two-body abrasive wear. The results will be published soon.

4 CONCLUSIONS

The following conclusions can be drawn from this investigation into the rôle of electroless nickel interlayer in the hardness and adhesion of TiN coatings deposited onto mild steel using a PVD method.

1. Electroless nickel (EN) can be successfully applied as a load-bearing interlayer on mild steel prior to TiN coating.
2. EN interlayers improve tribological behaviour of TiN coated mild steel.
3. Hardness increases with the thickness of EN interlayer.
4. The critical load to failure in scratch testing (Lc) increases with the thickness of EN interlayer. Lc is an indicator of adhesion and load-bearing capacity.
5. Harder substrates with thinner interlayers have the same effects as softer substrates with thicker interlayers.
6. Friction coefficient decreases with the thickness of EN interlayer.
7. The friction coefficient of the quenched substrates is lower than for as-received substrates.

References

1. C. Subramanian, K.N. Strafford, T.P. Wilks, L.P. Ward and K. Donaldson, Proc. Int. Tribology Conf. (AUSTRIB '94), 5-8 December 1994, Perth, Australia, p. 335.
2. D. S. Rickerby, S. J. Bull, T. Robertson and A. Hendry, *Surf. Coat. Technol.*, 1990, **41**, 63.
3. Y. I. Chen and J. G. Duh, *Surf. Coat. Technol.*, 1991, **46**, 371.
4. Y. I. Chen and J. G. Duh, *Surf. Coat. Technol.*, 1991, **48**, 163.
5. J. L. He and M. H. Hon, *Surf. Coat. Technol.*, 1992, **53**, 87.
6. J. C. Doong and J. G. Duh, *Surf. Coat. Technol.*, 1993, **58**, 19.
7. D. T. Gawe and U. Ma, *Mater. Sci. Technol.*, 1987, **3**, 228.
8. ASM Handbook, Vol. 5: 'Surface Cleaning, Finishing and Coating', 9th ed., 1982, p. 219.
9. C. Subramanian, K. N. Strafford, T.P. Wilks, L.P. Ward and W. McMillan, *Surf. Coat. Technol.*, 1993, **62**, 529.

1.3.6
Tribological-Structure Relationships of Alumina-Based, Plasma Sprayed Coatings

L.C. Erickson & T. Troczynski[1] and H.M. Hawthorne[2]

[1]METALS & MATERIALS ENGINEERING, UNIVERSITY OF BC, VANCOUVER, BC, CANADA

[2]SURFACE TECHNOLOGIES/TRIBOLOGY, NRC INSTITUTE FOR SENSORS & CONTROL TECHNOLOGY, VANCOUVER, BC, CANADA

1 INTRODUCTION

Ceramic plasma spray (PS) coatings are used both for protecting equipment components from degradation due to wear and corrosion, and as high temperature thermal barriers. For the former, in particular, coating structural integrity is critical. The microstructure of coatings that are deposited by atmospheric spraying typically exhibits a morphology of solidified droplets, or splats, and multiple pores and trapped inclusions, mostly at the inter-splat boundaries. Microscopic and, sometimes macroscopic, cracks are usually present also, as a result of local variations in temperature and thermal expansivity. Such microstructural features diminish coating integrity and, consequently, the wear and corrosion resistance of ceramic PS coatings is significantly lower than that of the corresponding bulk ceramics[1].

Aiming to improve their wear, corrosion and other properties by reduction of porosity and microcracks, surface modification of ceramic PS coatings has been attempted, with various degrees of success. Thus, laser treatment tends to be severe, with thermal shock and solidification stresses giving rise to surface cracking[2-4]. Sol-gel infiltration and subsequent heat treatment has been more successful[5,6]. This paper reports the results of scanning high energy arc lamp processing of inorganic sol-gel infiltrated ceramic PS coatings. Resulting coating microstructural and property changes were characterized by fractography, SEM/EDX and XRD analysis, depth sensing indentation, and abrasion and erosion tests.

2 EXPERIMENTAL

2.1 Materials

Alumina-titania coatings 260–300µm thick and with 5% porosity, were sprayed commercially by axial-feed plasma spray technology * in a mixed argon, hydrogen and nitrogen gas plasma, onto

---

* Axial IIITM torch by Northwest Mettech Corp., Richmond, B.C.
† Courtesy of Vortek Industries Ltd., Vancouver, B.C.
Micro-Abrasion

mild steel substrates that had been grit blasted with alumina. The precursor was Al_2O_3–13%TiO_2 (Norton 106 powder), particle size 15–45μm.

2.2 Coating Post-Treatment

Some coatings were post-treated by heat treatment alone, and some by infiltration plus heat treatment. Specimens were infiltrated by immersing in a solution of prehydrolyzed ethyl silicate containing 25% SiO_2 (Remet Corp. R–25 product) for 19 hours, followed by 20 min oven drying at 65°C. Heat treatment was performed by scanning the test samples under a high intensity arc lamp† at a speed of 25.4 mm/sec. Low (LL) and high levels (HL) of heat flux were used at 700 W/cm^2 and 1400 W/cm^2, respectively.

2.3 Coating Characterization

2.3.1 EDX & XRD Analyses. Information on elemental and phase composition of as-deposited and treated coatings was obtained from energy dispersive X-ray analysis in the SEM and X-ray diffraction using Cu-K$_\alpha$ radiation in a Siemens D5000 powder diffractometer.

2.3.2 Microstructural and Morphological Characterization. Coating microstructures were examined by optical microscopy, while scanning electron microscopy (SEM) was used to characterize both the morphology of surfaces before and after wear testing, and fracture surfaces of the different coatings. The latter were prepared by cutting a notch almost through the substrate, cooling the specimens in liquid nitrogen, and breaking them with the coatings in tension.

2.3.3 Depth-Sensing/Indentation. Depth sensing indentations (DSI) in the load range 0.1–1N were made on both polished cross-sections and surfaces ground parallel to the top (planar sections), using a Fischerscope H 100B instrument with a Vickers indentor[7]. Elastic (Young's modulus, recovery) and plastic (hardness) properties of the coating were determined from the load vs. indentation depth data on loading and unloading. Standard Vickers microhardness measurements were also made on cross-sections and some planar sections at a load of 3N.

A standard dimple grinder for preparing TEM samples (Dimpler model D–500i) was used for micro-abrasive wear testing [8], as illustrated in Figure 1. A stainless steel grinding wheel (15 mm diameter, 2 mm wide) with a rounded edge, rotated at 280 rpm (0.22 m/sec. sliding velocity) against the slowly rotating specimen under a load of 0.1N in a water-based, 3μm particle size diamond slurry. The coating surfaces were ground flat and finish polished with 1μm diamond paste prior to testing. Wear scar depth was monitored continuously during testing and total wear volume was checked by surface mapping profilometry after testing.

2.3.4 Erosion. The comparative resistance to impact erosion of specimen surfaces was carried out in a modified sand-blaster. An air-entrained jet of 60 grit angular alumina particles at approximately 70 m/s and 55 g/min flux was directed at 90° nominal impact angle against the coatings. Eight specimens were mounted into the urethane rubber surface of a turntable such that their top surfaces only were exposed to the erodent stream as they passed under the jet. The jet nozzle was positioned 1 cm away from the samples. Erosion resistance of post-deposition treated coatings relative to that of as-deposited ones was determined in these preliminary experiments from both weight loss and profilometric measurement of wear scar volumes. The weight loss was obtained after running the test for 5 minutes, during which each specimen was exposed to the erodent stream for a total of 7.5 seconds.

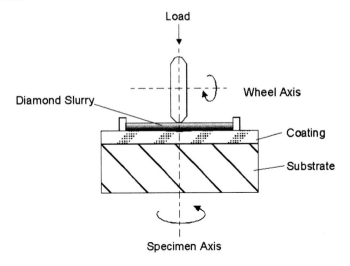

Figure 1 *Schematic illustration of micro-abrasion test*

3 RESULTS AND DISCUSSION

3.1 Microstructural Characterization and Properties

X-ray diffraction shows that the as-deposited coatings consisted mainly of γ-alumina, with a small amount of α-alumina, and TiO_2. The metastable γ-phase is formed by rapid cooling of the splats during plasma spraying[9], while the α-phase is usually attributed to unmelted particles in the coating. From the approximate relative amounts of the different phases found in treated samples, Table 1, the high level heat-treatment, especially, induced a phase change to the more stable, and harder, α-Al_2O_3 form.

3.2 Fractography

Fractography can help elucidate the structural changes induced by the various coating treatments. Transverse fracture surfaces show a distinct splat morphology, with average splat thickness of

Table 1 *Phase composition of coatings based on XRD results*

Sample			Extent of phases present				
Infiltrated	Heat-treated	Designation	α–Al_2O_3	γ–Al_2O_3	TiO_2	Al_2TiO_5	SiO_2
As-sprayed	As-sprayed	Sprayed	*	****	*		
No	Low level	LL	**	***	*		
Yes	Low level	I/LL	**	***			*
No	High level	HL	****			**	
Yes	High level	I/HL	****			**	

4–6μm, throughout the as-deposited coating, Figure 2. There appear to be more inter-splat delaminations close to the substrate. Low level heat-treatment alone (LL) produced no significant microstructural change. With infiltration and low level heat-treatment (I/LL), the splat morphology was obscured and porosity reduced in the top third of the coating, Figure 3 (a). However, the greatest microstructural changes were induced by the high level heat treatment, with (I/HL) or without (HL) infiltration. Figures 3 (b) and 3 (c) show extensive recrystallization in the top portion of the HL and I/HL coatings, again obscuring any sign of splats, such that the microstructure now looks more like that of a monolithic ceramic. This results from the phase change from γ- to α-alumina, which was also evidenced by the coating colour change noted previously[10].

3.3 Indentation Hardness

Figure 4 illustrates typical results obtained from indentation probing the cross-sections of various post-deposition treated coatings at the two load extremes used in depth sensing indentation. The hardness values are those calculated from the plastic work (HV2) and the volume of the residual indentation[11]. Though not as pronounced as for chromia/titania coatings[7], the tendency towards higher hardness from substrate to outer surface may result from both thermal gradient effects during spraying and coating treatment effects. The low load hardness values reflect the microstructural periodicity observed with indentations of similar or smaller size to average splat thickness on ceramic thermal spray coatings[7].

Since we are mainly interested in effects on coating wear of post-deposition treatments which proceed from their top surface, hardness results from the top third of the coatings are summarized in Table 2. Included are HV_2 and corresponding modulus and recovery values from DSI, along with hardness–under–load values (HV_L) obtained directly from the Fischerscope software, and standard Vickers microhardness (HV_1) results. The data in Table 2 show that, with a few exceptions, treatments improve hardness and modulus, from about 15 % on low level heat treatment alone to about 30 % on combined infiltration and high level heat treatment. On the other hand, the treatments have no significant effect on depth recovery. Both this, and the fact that trends are broadly similar for all three measures of hardness (when HV_L is derived from the combined plastic and elastic deformation during indentation while HV_1 and HV_2 are from plastic deformation only) indicate that elastic and plastic contributions to the deformation remain substantially unchanged by the treatments.

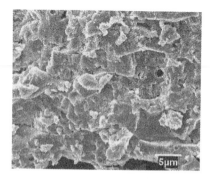

a   b

Figure 2 *SEM micrographs of as-sprayed fracture surfaces: (a) at surface,*
(b) near substrate

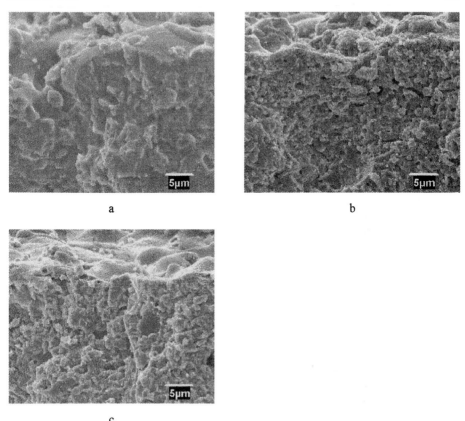

a   b

c

Figure 3 *SEM micrographs of fracture surfaces: (a) I/LL, (b) HL and (c) IHL coatings*

HV$_2$ measurements are intrinsically free from indentation size effects[11]. Therefore, the load dependence seen in the present results may reflect differences in coating (cross-section) response to probing at different scales, with inter-splat bonding effects influencing the larger load results. At 3N, 1N and 0.1N loads, indentations span about five, three and one splat thicknesses, respectively. The lower measured hardness at higher load would then point to inter-splat bonding somewhat weaker than the cohesion within splats. However, further work is needed to establish this more clearly.

3.4 Micro-Abrasion

Cumulative wear plots for coating specimens showed consistently smaller multiple-splat scars for infiltrated and heat-treated coatings, at least to 60–80 μm below polished top surfaces[10]. Figure 5 shows that the corresponding steady-state wear rates correlated inversely with the Vickers microhardness measured on the top surfaces of the same coatings. The dimple scar surfaces on the harder, treated coatings showed fewer individual scratches after this mild abrasion. These coatings were also somewhat less susceptible to grain (splat) pull-out during the more severe abrasion of metallographic polishing of specimens[10]. The latter would suggest a strengthening of the inter-splat bonding in treated specimens but these qualitative observations require verification.

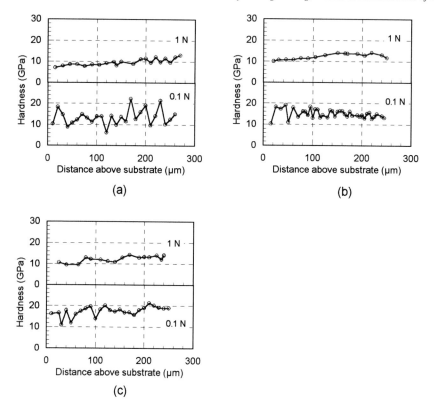

Figure 4 *DSI results at two loads on cross-sections of (a) as-sprayd, (b) I/LL and (c) I/HL coatings*

Table 2 *Indentation results on top third of cross-sections of various coatings*

Parameter	Load (N)	Sprayed (GPa)	LL (GPa)	I/LL (GPa)	I/HL (GPa)
HV_1	3	11.5	11.6	14.7	17
HV_2	1	11	12.8	13.48	13.46
HV_L	1	7.7	8.8	9.4	9.7
$E/(1-\upsilon^2)$	1	180.5	203	213	236
Recovery	1	71.8%	70.5%	70.9%	71%
HV_2	0.1	14.9	16.8	14.8	18.55
HV_L	0.1	13.4	14.4	14.3	16.6
$E/(1-\upsilon^2)$	0.1	198	239	255	276
Recovery	0.1	67.6%	65.5%	69.2%	67%

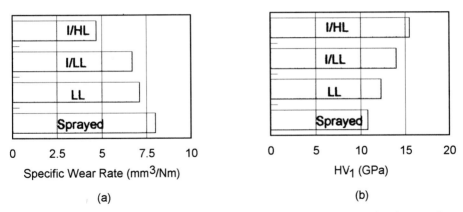

Figure 5 *Comparison of micro-abrasive wear, (a) with Vickers microhardness and (b) measured on the top surfaces of some plasma sprayed coatings*

3.5 Erosion

Results from preliminary (duplicate) experiments, Figure 6, show that erosive wear is also reduced by the post-deposition treatments, the high temperature, infiltrated (I/HL) coating showing the best erosion resistance. SEM examination of worn surfaces provided some evidence that inter-splat cracking and spalling in the as-sprayed specimens was reduced in the treated coatings, but a quantitative measure of this remains to be established by, e.g., image analysis. Similar erosion test results were interpreted in terms of inter-splat bond strengths of plasma sprayed coatings[12].

4 CONCLUDING REMARKS

Post-deposition treatments impart better abrasive and erosive wear resistance to the alumina-based thermal spray coatings, mainly through improvement of their microstructure and mechanical properties. The microstructural refinement resulting from infiltration and heat treatment provides greater coating integrity with fewer pores and micro-cracks and, especially with the higher temperature processing,

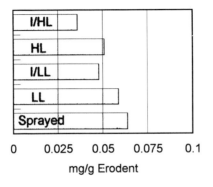

Figure 6 *Erosive wear results*

a harder phase composition with a somewhat less splat-like morphology. These microstructural improvements are reflected in greater resistance of the coatings both to quasi-static penetration of sharp indenters and surface damage by hard particles in low stress abrasion and impact erosion tests. Qualitative indications of inter-splat bond strength enhancement from the present experiments need to be established more firmly.

Acknowlegements

Thanks are expressed to D. Ross and H. Tai of Northwest Mettech Corp., for provision of coatings, and to S.Yick for assistance with preparation of some of the figures.

References

1. K. Furukubo, S. Oki and S. Gohda, "Relationship Between Wear and Microstructures of Ceramic Spray Coatings", International Advances in Coatings Technology, The Thermal Spray Division of ASM International, Metals Park, OH, USA, 1992, p. 705.
2. R. Sivakumar and B. L. Mordike, *Surface Engineering*, 1988, **4**, 127.
3. J. M. Cuetos, E. Fernandez, R. Vijande, A. Rincon and M. C. Perez, *Wear*, 1993, **169**, 173.
4. A. Wang, Z. Tao, B. Zhu, J. Fu, X. Ma, S. Deng and X. Cheng, *Surface and Coatings Technology*, 1992, **52**, 141.
5. K. Niemi, P. Sorsa, P. Vuoristo and T. Mantyla, "Thermally Sprayed Alumina Coatings with Strongly Improved Wear and Corrosion Resistance", Thermal Spray Industrial Applications, C.C. Berndt and S. Sampath, Eds., Thermal Spray Division of ASM International, Metals Park, OH, USA, 1994, p. 533.
6. I. Berezin, and T. Troczynski , *J. Materials Science Letters*, 1996, **15**, 214.
7. L. C. Erickson, T. Troczynski and H. Hawthorne, Proc. International Thermal Spray Conference, Kobe, Japan, May 1995, 743.
8. A. Kassman, S. Jacobson, L. Erickson, P. Hedenqvist and M. Olsson, *Surface Coatings and Technology*, 1991, **50**, 75.
9. S. Safai and H. Herman, in Treatise on Materials Science & Technology, Academic Press, 1981, **20**, 183.
10. L. C. Erickson, H. M. Hawthorne, T. Troczynski, S. Hogmark and M. Olsson, 'Advanced Ceramics for Structural and Tribological Applications' (Eds. H. M. Hawthorne and T.Troczynski) Met. Soc. of CIM, Montreal, 1995, p. 557.
11a. R. Berriche and R.T. Holt, in 'Surface Engineering Volume II: Engineering Applications' (Eds. P.K. Datta & J.S. Gray), Royal Soc. Chem., 1993, p. 292.
11b. R. Berriche, *Scripta Metallurgica et Materialia*, 1995, **32**, 617.
12. A. Ohmori and C. J. Li, in 'Plasma Spraying – Theory and Applications' (Ed. S. Suryanarayanan) World Scientific, Singapore, 1993, p. 179.

1.3.7

The Tribological Characteristics of a Detonation Gun Coating of Tungsten Carbide under High Stress Abrasion

M.L. Binfield and T.S. Eyre

BRUNEL UNIVERSITY, UXBRIDGE, MIDDX, UB8 3PH, UK

1 INTRODUCTION

Parallel gate valves are predominantly used to control flow in subsea pipelines used for the production of oil and gas, where demands are high in terms of reliability and freedom from maintenance. Their role will become of increasing importance as the search for oil progresses to more remote locations where retrieval costs are at present uneconomic. Therefore, it is essential to fully understand the tribological characteristics of these valves which have been allocated a high research priority. As well as being located in remote areas, gate valves also have to experience very harsh and variable service conditions. Typically these might comprise:

- downhole temperatures of 110–150°C,
- the presence of hard sand particles, typically 50 ppm,
- variable sand size but typically 70 µm,
- variable fluid velocity (approximately 5 ms⁻¹ for long distance oil pipelines),
- the velocity through the valve will change through an opening/closing cycle,
- the presence of corrosive agents such as sulphur and water in the crude, and
- infrequent valve operation.

The principle of operation of the gate valve is very straightforward and is shown schematically in Figure 1. Predominantly, sealing is accomplished on the downstream side of the valve with the pressure of the processing liquid forcing the gate onto the seat. Therefore, the integrity of the sealing surfaces of the valve is critical to its performance and historically it is this area in which problems occur. Examination of worn test valves has shown that the presence of sand contributes the majority of damage to the sealing surfaces, be it by erosion or three body abrasion[1].

The primary method of wear prevention to the sealing interface has concentrated on the use of hard surface coatings and generally it has been found that harder materials such as stellite, tungsten carbide and ceramic have provided the greatest resistance. Tungsten carbide LW45 is a coating designed specifically for this application and is currently the most popular. It is deposited by detonation gun (D–Gun) to a thickness of approximately 100µm.

It is useful before proceeding further to summarize the detonation gun technique, which is reviewed in greater detail elsewhere[2–5]. The process was patented and developed by the Union

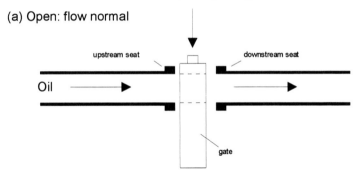

(a) Open: flow normal

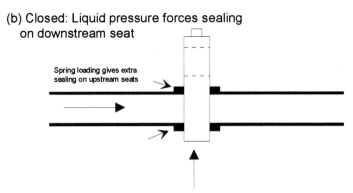

(b) Closed: Liquid pressure forces sealing on downstream seat

Figure 1 *Schematic of valve operation*

Carbide Corporation[6] and is currently available exclusively from Praxair Surface Technologies. Porosity is amongst the lowest for thermally sprayed coatings (½ to 1%) and the reported coating substrate bond strength for tungsten carbide-cobalt coatings of above 170 MPa is more than twice that reported for the corresponding plasma sprayed coatings. However, on the downside, the process is expensive compared to other thermal spraying techniques[7], caused by the relatively high cost of installation and low spray rate[3].

The detonation gun consists of a water-cooled barrel approximately 1.4 m long with an inside diameter of 25 mm closed at one end, Figure 2. At the closed end is a sparking plug and a system of valves through which are metered the appropriate quantities of acetylene, oxygen and the powder to be sprayed. The gas mixture is ignited either 4.3 or 8.6 times per second (generally the latter), and a short period of burning or combustion ensues. The flame front accelerates, compressing and heating the gas zone immediately ahead of it and, depending upon the gas mixture being compressed, a critical temperature is reached where self-ignition of the gas occurs, producing a detonation or shockwave. This shockwave compresses and raises the gas temperature inducing chemical reaction of the gas mixture, producing gas temperatures greater than 3,000°C. If the powder particles are suspended in the gas mixture prior to detonation, the rapid expansion of the reacted gases forces the particles from the nozzle at approximately 800 ms⁻¹. After the powder has exited the barrel, it is purged by a pulse of nitrogen. The repeated detonation cycle produces a coating structure which is built up of a series of detonations or pops on the prepared substrate which is placed 50–100mm in front of the barrel.

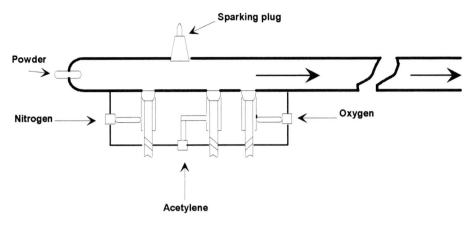

Figure 2 *Schematic representation of the detonation gun*

The abrasive wear performance of detonation gun coatings of tungsten carbide have been assessed before but the data has been in the form of a ranking which has shown the comparative performance of the coatings. Tests were conducted to ASTM G65 (rubber wheel abrasion[8]) or using a pin-on-disc geometry with a sandstone disc[9]. In both instances, the detonation gun coatings have given a good performance, but there has been no attempt to examine the underlying failure mechanisms. This paper will therefore, review a laboratory-based experimental programme aimed at understanding the abrasive wear failure mechanisms of LW45 and, in doing so, it will be possible to suggest methods by which its wear performance may be enhanced.

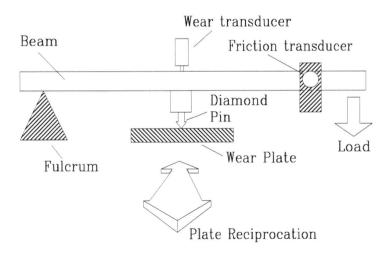

Figure 3 *Schematic representation of the reciprocating diamond-on-plate test*

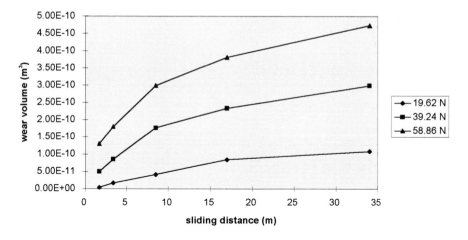

Figure 4 *Wear volume vs sliding distance*

2 EXPERIMENTAL PROCEDURE

To simulate the high stress abrasive wear performance of detonation gun, LW45 coatings, a test that had previously been developed at Brunel University was used[10]. The technique is relatively straightforward and utilizes a reciprocating motion with a Rockwell diamond indenter as a pin (Figure 3). Loads of 19.62, 39.24 and 58.86 N were used at an average speed of 0.027 ms^{-1} (41 rpm). The track length was 20 mm and total sliding distance was 8.2, 16.4 and 32.8 m respectively. Two materials were evaluated; LW45 and for comparison AISI 410 stainless steel. LW45 samples were coated by Praxair Surface Technologies, Swindon to a thickness of 100 µm. Test surfaces were lapped to a finish of 0.05µm R$_a$. No attempt was made to control the test environment but it was measured as $20 \pm 3°C$ and a relative humidity of 55–70%. Wear was measured by profilometer traces across the wear scar and converted into wear volume. Post test examination of the worn surface was conducted using a scanning electron microscope (SEM).

3 RESULTS

The wear data generated during this series of tests is presented in Figures 4–6 and gives both wear volume and specific wear rate (SWR) as a function of load and sliding distance. The latter value is wear volume divided by both load and sliding distance and helps to show whether there are any underlying changes in wear rate occurring at any given point. If the same wear mechanism was prevalent throughout, then the resultant SWRs would be expected to be identical regardless of load or sliding distance.

Figure 4 shows the volume of material removed during reciprocating diamond-on-plate tests conducted on LW45 coatings under three loads (19.62 N, 39.24 N and 58.86 N). It can be seen that wear volume increases as a result of sliding distance and load. For the two highest loads there is a period of higher running-in wear before the wear rate stabilises.

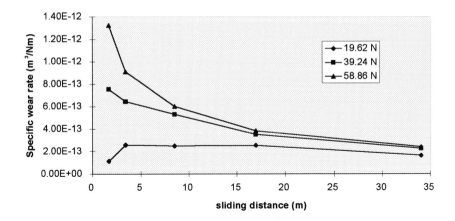

Figure 5 *Specific wear rate vs sliding distance for LW45*

Figure 5 shows specific wear rate as a function of sliding distance for three loads. It can be seen that the SWR decreases with sliding distance for the highest two loads and remains constant for the lowest load. However, the SWR is approximately equal for each load and, therefore, it is reasonable to assume that it is independent of load for this coating. This implies that the same failure mechanism occurs in all tests.

Figure 6 shows the comparative performance of LW45 and the AISI 410 substrate material. Note the logarithmic scale on the y-axis which clearly indicates that the tungsten carbide coating gives approximately two orders of magnitude less wear than the stainless steel.

Taper sections of all of the wear tracks were made[5] and, as well as providing an increase in magnification, the technique also allows the surface and the subsurface to be viewed simultaneously. Figure 7 shows the LW45 surface after it has worn under a load of 19.62N for

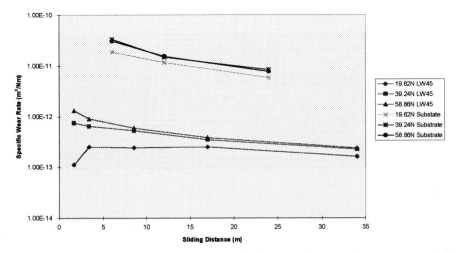

Figure 6 *Wear rate of LW45 in comparison to AISI 410 stainless steel substrate material*

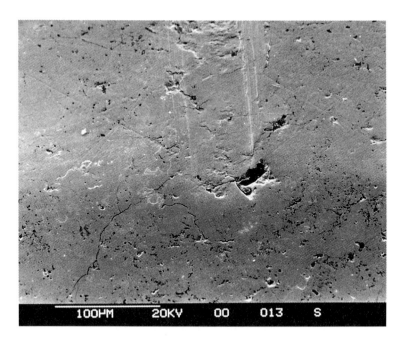

Figure 7 *Scanning electron micrograph of small wear scar (19.62N, 1.69m)*

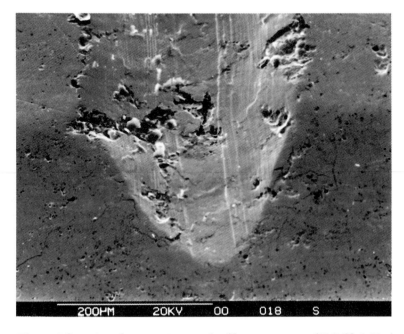

Figure 8 *Scanning electron micrograph of large wear scar (58.86N, 1.69m)*

Figure 9 *Scanning electron micrograph of edge of wear scar (58.86N, 1.69m)*

a sliding distance of 1.69m. A diamond shaped indent that has been made in the coatings can be seen alongside abrasive marks in the direction of sliding. The wear tracks also show evidence of large fragments of material breaking off especially around the edges of the scar. There is little information to be obtained by examination of the microstructure underneath the wear track. Microhardness readings were taken immediately below the wear track, however the values were similar to those obtained on an untested section and, therefore, no work hardening could have taken place.

Figure 8 shows a test sample subjected to a load of 58.86N for a sliding distance of 1.69m. The scar is significantly larger than that shown in Figure 7 (note the different scale) which was run under a lower load and also, on this occasion, further information upon the underlying failure mechanism may be obtained by closer examination.

Figure 9 gives a higher magnification of the edge of the wear track where it can be seen that the coating has been extruded out of the track resulting in cracking below this area. Wear by plastic deformation, which results in strains above the strain-to-failure ratio, therefore cracking the material, leads to extremely large material removal rates, and consequently high wear rates. It was also noticeable that material removal tends to occur predominantly at the wear track edges under all conditions.

Figure 10 shows a higher magnification of a different high load scar, this time at the base of the wear track. Fine abrasive marks can be seen on the sides of the track possibly due to debris entrapment. At the base itself, delamination type cracks are visible transverse to the direction of sliding. Crack interlinking may be seen midway up the wear scar in Figure 11.

4 DISCUSSION

Failure analysis has clearly shown three-body abrasion caused by the ingress of sand particles to be of primary importance. The reciprocating diamond-on-plate test was chosen as one that simulated this mechanism and has the advantage that it represents a single point contact and, therefore, does not experience problems with misalignment. Furthermore, analysis is simplified in comparison to testing in the presence of sand particles in that there will be no element of rolling, a known attack angle and a known counterface angularity.

It has been seen that there is a clear correlation between wear volume and load and also between wear volume and sliding distance (Figure 4). Furthermore, after approximately 17m, the specific wear rates converge and may be taken as being independent of load and therefore, it is proposed that the same material removal mechanism predominates at all of the loads tested.

This assumption is supported by reference to the scanning electron micrographs taken from taper sections through the scar (Figures 7 and 8) where essentially, the scar appearance is the same regardless of load or sliding distance. It is proposed that there are two fundamental and interrelated failure modes for the coating under this test. These mechanisms are related to the two principle types of abrasive wear; namely plastic deformation and brittle fracture.

Probably the most appropriate mode of failure to consider for a hard coating such as LW45 is brittle fracture. Several models have been proposed for brittle failure[11,12], and are summarised by Hutchings[13]. Lawn and Swain[12] have shown that the application of a sharp indenter above a certain load for a brittle material will lead to the creation of a median vent crack in the subsurface area. Upon relaxation of this strain, lateral cracking will occur in the vicinity of the deformed region. These lateral cracks can lead to the free surface and result in material removal.

Figure 10 *Scanning electron micrograph of base of wear scar (58.86N, 33.8m)*

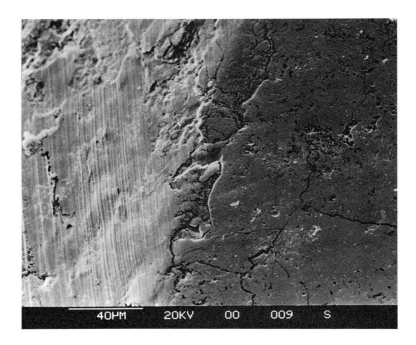

Figure 11 *Scanning electron micrograph, midway up wear scar (58.86N, 33.8m)*

There is an increased tendency for this to happen in a coating such as LW45 due to the presence of microcracks throughout the microstructure, shown clearly in Figure 12. The coating is unable to accommodate the strain by plastically deforming alone and will, therefore, fracture at its weakest point which corresponds to subsurface flaws. The surface produced by such a mechanism can be seen in the centre of the wear track, where compressive stresses are greatest, and is shown in Figure 10. No plastic deformation-dominated abrasion is apparent and wear will ensue by the removal of platelet-type debris initiated by lateral cracking. A similar mechanism can be seen down the sides of the scar, although the compressive stresses in this region will be less. The mechanism for brittle fracture-dominated abrasion of LW45 is illustrated schematically in Figure 13.

The coating does not wear in a uniformly brittle manner, however, and examination of the edge of the wear track in Figure 9 shows a representative scar worn at 58.86N load for 1.69m. As well as being exposed to compressive stresses in this area the coating will also be subject to a shear stress which will force coating material to the surface. This results in plastic deformation and leads to material being extruded over the edge of the wear track. However, the strain-to-failure of the coating is low and therefore, the deformed material will subsequently fracture and result in material removal. Such a mechanism is shown schematically in Fig. 14.

It can, therefore, be seen that abrasive wear experienced by LW45 in this test is not represented solely by brittle fracture or plastic deformation. However, examination of the micrographs of the scars (Figures 7–11), shows that the majority of wear occurs at the wear scar edges where shear stresses and plastic deformation are dominant.

Regardless of the type of brittle failure, it is suggested that the elimination of microcracks will lead to a significant reduction in wear. This is supported by the work of Boas and

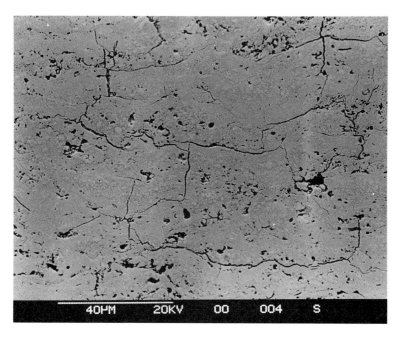

Figure 12 *Microstructure of LW45*

Bamberger[14] who determined the low load abrasive wear characteristics of plasma sprayed coatings. They suggested that the existence of a high defect level in a hard material will promote crack propagation in a subsurface where stress is applied. Ductile materials with a lower flaw level, in contrast, would exhibit slow crack propagation and even halt cracks. Their experimental work found a marked difference in the wear behaviour of a virgin plasma coating and one that had been subjected to a subsequent laser melting operation which removed faults inherent in the coated microstructure. Boas and Bamberger also proposed a delamination type failure , similar to the one proposed for brittle failure in this paper, resulting in removal of platelet material from the wear scar. Further support for this theory has been gained by the abrasive wear testing of Super D-Gun™ coatings which show freedom from microcracking, where the wear rate has been reduced by at least 50%[8].

As there is no change in wear mechanism with load or sliding distance, it is proposed that for this thickness of coating (approximately 100 μm), the hardness of the substrate plays no part in the failure mechanism. There is no evidence that deformation in the coating may be attributable to substrate collapse.

5 TEST LIMITATIONS

The reciprocating diamond-on-plate test reported in this study is designed to simulate the high stress abrasion that is likely to occur due to the ingress of foreign hard particles. Nevertheless, it should be remembered that in a real life application, it is unlikely that a particle would continually slide against a particular area (unless, of course, it had become embedded in the counterface). Furthermore, it would not be as hard or as large as the Rockwell diamond,

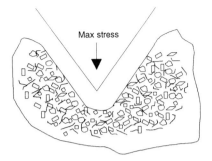

(a) Coating under maximum stress at centre of wear track

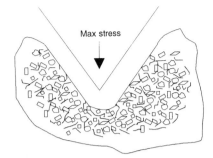

(b) Existing microcracks increase in size under stress

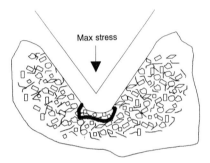

(c) Cracks interlink and spread to free surface

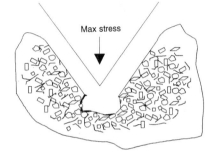

(d) Platelet material removed

Figure 13 *Schematic representation of brittle fracture-dominated abrasion*

would probably not possess such a sharp incident angle and would also be likely to experience some rolling element in its motion.

Perhaps the greatest deviance from service conditions is the fact that diamond has a hardness of 6,000–10,000H_v in comparison to that for silica of around 800H_v. The particle hardness has been shown to have a profound effect on the abrasion of all materials, and it is known that unless the ratio of hardnesses of the abrasive, H_a to the surface H_s exceeds 1.25 then no plastic deformation would be expected[13]. Therefore, as LW45 has a bulk hardness of around 900 H_v, it can be seen that it would normally be expected to resist abrasion by silica particles. Consequently, the regime generated by this test may not be strictly applicable to the application.

Furthermore, the size of the abrasive particle has been shown to be of significance in the abrasive wear performance of sintered cermets and may also be appropriate for sprayed carbides. If the test scenario is similar to the one described here, where the dimension of indentation is substantially greater than the size of the carbide particles and their separation, then the material will behave in a similar manner to a homogeneous solid. If, on the other hand, the hard phase particles are comparable in size to the indentation then the material will behave heterogeneously (Figure 15). For the vast majority of abrasive particles encountered in petrochemical applications, the size of the particle (50–200µm) will be large when compared to the size of the carbide particles (1–2µm)[5]. Such a scenario is replicated by the diamond-on-plate test method and in such circumstances, the material will behave homogenously and give the best

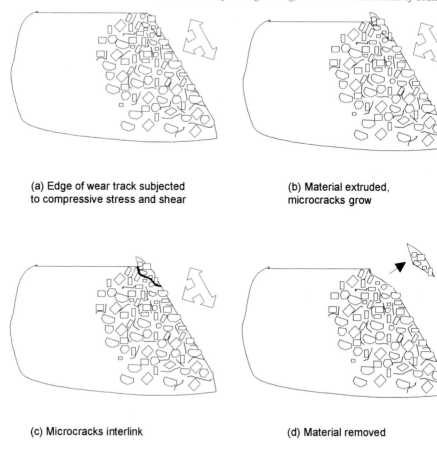

(a) Edge of wear track subjected
to compressive stress and shear

(b) Material extruded,
microcracks grow

(c) Microcracks interlink

(d) Material removed

Figure 14 *Schematic representation of plastic-dominated abrasion*

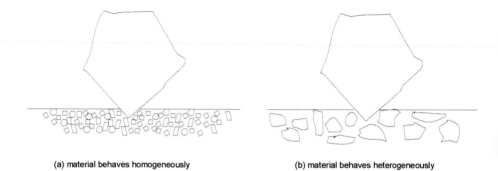

(a) material behaves homogeneously

(b) material behaves heterogeneously

Figure 15 *Influence of carbide grain size on abrasion*

wear resistance. However, if a great quantity of smaller abrasive particles was present, whose size was of the same order as the carbide particles, then the coating would behave heterogeneously and preferential binder removal could occur.

The test represents a scenario that is likely to far outstrip, in terms of severity, the conditions that are encountered in service. Consequently, it is not possible to comment on potential service lifetimes from the data produced. Nevertheless, for the same reasons mentioned above, the particle variables such as hardness, angularity and size will be completely removed from the test and the H_s/H_a value can be discounted for all engineering materials. Therefore, regardless of the provisions already stated relating to the suitability of the diamond–on–plate test in simulating this particular set of operating conditions, the test should prove excellent at producing comparative material data.

6 CONCLUSIONS

1. Detonation gun coatings of tungsten carbide show a significant improvement in tribological performance when compared to stainless steel substrate material.
2. Wear volume increases as a function of both load and sliding distance.
3. The same wear mechanisms are prevalent under all conditions tested and correspond to a combination of brittle fracture and plastic deformation.
4. The reduction in the amount of inherent microcracking within the structure of the detonation gun coating should lead to a significant improvement in abrasive wear resistance.

References

1. M. L. Binfield , T. S. Eyre , F. A. Davis, 'Replication Techniques Within The Oil Industry', Tribologia, Vol. 11, No. 4, p. 56, 1992, 5th Nordic Symposium on Tribology, Helsinki 1992.
2. 'Putting A Hard Face On The World', published in Metallurgica, July 1983, p. 299.
3. R. C. Tucker, in 'Deposition Technologies for Films and Coatings, Developments and Applications', F. Rointan, Bunsah et al. eds., 1982, Noyes Publications, New Jersey, USA.
4. C. J. S. Guest, *Trans. Inst. of Metal Finishing*, 1986, **64**, 33.
5. M. L. Binfield, Ph.D. Thesis, Brunel University, Uxbridge, U.K., 1995.
6. R. M. Poorman, H. B. Sargent and H. Lamprey, U.S. Patent 2,714,553, August 2, 1955.
7. M. L. Thorpe, *Adv. Mats. and Processes*, 1993, **143**, 50.
8. M. Watson, A. Davies, J. Eaves, K. Ferguson, 'CAST Group 2, High Stress Abrasive Corrosive Wear', Final Report 1991, National Centre of Tribology, Warrington, UK.
9. F. A. Davis, D. Cooper, R. J. K. Wood, *Tribologia*, 1992, **11**, 107.
10. R. Nogueira, PhD Thesis, Brunel University, 1992.
11. Evans A.G., 'Abrasive wear in ceramics in an assessment', NBS Spec. Publ. 562, 1979, p. 1 (National Bureau of Standards, Washington DC).
12 B. R. Lawn and M. V. Swain, *J. Mat. Sci.*, 1975, **10**, 113.
13. I. M. Hutchings, 'Tribology, Friction and Wear of Engineering Materials', Edward Arnold, 1992.

14.   M. Boas, M. Bamberger, *Wear*, 1988, **126**, 197.
15.   N. P. Suh, *Wear*, 1977, **44**, 161.
16.   I. M. Hutchings, Proc. 2nd European Conf. on Advanced Materials and Processes, Vol. 2, Institute of Materials, p. 56, 1992.

1.3.8

The Influence of the Substrate Bias Voltage on the Adhesion of Sputtered Nb Coatings and Correlation with the Hardness Data

L. P. Ward[1] and P. K. Datta[2]

[1]DEPARTMENT OF CHEMICAL AND METALLURGICAL ENGINEERING, ROYAL MELBOURNE INSTITUTE OF TECHNOLOGY, CITY CAMPUS, PO BOX 2476V, MELBOURNE 3001, AUSTRALIA

[2] SURFACE ENGINEERING RESEARCH GROUP, SCHOOL OF ENGINEERING, UNIVERSITY OF NORTHUMBRIA AT NEWCASTLE, UK

1 INTRODUCTION

In the field of biomaterials, Nb and Ta metals have attracted attention because of their excellent corrosion resistance and biocompatibility. However, as both Nb and Ta are mechanically weak, they cannot be used in pure bulk form for the construction of load-bearing implants. To overcome this problem, consideration has been given to using these metals as coatings deposited on appropriate substrates. Nb and Ta coatings deposited on 316L stainless steel substrates using the ion-plating process have been shown to suffer no loss in corrosion resistance and biocompatibility[1-4].

However, the most important property that needs attention in any coating system is the coating/substrate adhesion. This property, like corrosion resistance and biocompatibility needs to be correlated to the morphology and structure of the coatings which, in turn, are controlled by the process deposition parameters.

Previous studies by Ward and Datta[5] have shown that the adhesion and hardness of Nb coatings were influenced by the film thickness and to a lesser extent, the substrate surface finish. This paper reports on the film-to-substrate adhesion and hardness of Nb and Ta coatings as a function of the substrate bias voltage and interpretation of the results in terms of the structure and morphology.

2 EXPERIMENTAL

The surface of 316L stainless steel specimens, having dimensions $25 \times 25 \times 3$ mm^3, were highly polished using various grades of SiC paper followed by 6μm and 1μm diamond paste. Prior to deposition, specimens were initially rinsed in paraffin, ultrasonically degreased for five minutes in the vapour of 1,1,1 trichloroethane, followed by further ultrasonic cleaning in "Decon 90" detergent for 5 minutes. After rinsing in water, the substrates were solvent dried in 1,2, dichlorodifluoroethane for 15 minutes.

All depositions were carried out in a Leybold-Hereaus L560 coating unit. The Nb sputter target was 150 mm in diameter and 6 mm thick The chamber was initially pumped down to 8×10^{-4} Pa to minimise any contamination effects due to residual oxygen and/or chamber contaminants, then back filled with argon gas to the desired value. Specimens were sputter cleaned for 30 mins at − 800 V substrate bias and 1.1 Pa argon gas pressure, prior to deposition.

The target was energized without interrupting the cleaning process. Prior to rotating the specimen into the coating flux, the target itself was sputter cleaned for a few minutes to remove any surface contaminants. Film thickness and deposition rate were monitored using a quartz crystal sensor, situated next to the substrate holder. The deposition rate was controlled by varying the power through the sputter target. No external substrate heating was employed during any of the coating cycles and the maximum substrate temperature reached was 200 to 230°C. This was measured by inserting a thermocouple in a dummy specimen, placed in the plasma adjacent to the sample to be coated. The substrate-to-target distance was kept constant at 60mm.

The scratch testing machine employed in the study here was a REL paint scratch tester, modified to incorporate a Rockwell "C" diamond indentor as the scratching stylus, in accordance with standard testing procedure[6]. Coated specimens were repeatedly passed under the loaded stylus at a constant speed of 11 mm/min. The load was increased with each pass from 100g to 2kg, in 100g intervals. The associated scratches, approximately 15 mm long and 1 mm apart, were examined using optical and scanning electron microscopy to define the onset of coating failure so that the critical load can be identified. Such examinations also revealed the nature of failure and mechanisms responsible for the loss of coating/substrate adhesion.

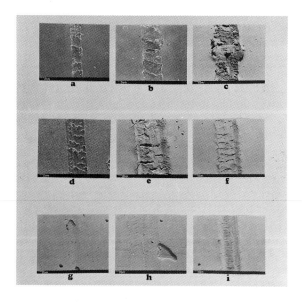

Figure *1 SEM micrographs showing scratch test results as a function of the substrate bias voltage;*
(a) 0 V, 100g; (b) 0V, 300g; (c) 0V, 1600g; (d) – 200V, 100g; (e) – 200V, 900g;
(f) – 200V, 2000g; (g) – 400V, 200g; (h) – 400V, 900g; (i) – 400V, 2000g

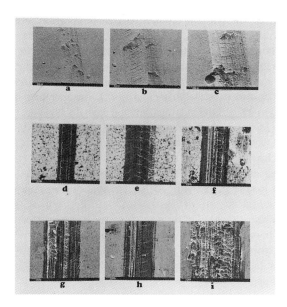

Figure 2 *SEM micrographs showing scratch test results for Nb coatings as a function of the substrate bias voltage; (a) – 600V, 400g; (b) – 600V, 1000g; (c) – 600V, 1600g; (d) – 1000V, 400g; (e) –1000V, 700 g; (f) – 1000V, 1900g; (g) – 2000V, 100g; (h) – 2000V, 300g; (i) – 2000V, 800g*

3 RESULTS

3.1 Scratch Test Results for Nb Coatings as a Function of the Substrate Bias Voltage

For the range of substrate bias voltages considered here, scanning electron micrographs of the scratch channels produced on Nb coatings at selected loads, are shown in Figure 1 and Figure 2. For coatings deposited at 0V substrate bias voltage, the scratch channels developed are shown in Figure 1 (a)–(c). At low loads of 100g, the coating displayed severe deformation, consisting of built-up ridges within the channel at regular intervals, as shown in Figure 1 (a). As the load was increased to 300g, further plastic deformation of the coating was observed, the coating being removed from the substrate within the scratch channel at regular intervals, shown in Figure1 (b). At high loads of 1600g, the scratch channel consisted of highly deformed coating material, which was picked up by the stylus and re-deposited within the scratch channel, as depicted in Figure 1 (c).

On the application of a bias voltage of –200V, the failure of the coating was observed to occur in a less ductile manner. At 100 g load, although the scratch channel consisted of ridges of plastically deformed material, cracks became visible at regular intervals along the channel length, indicating a more brittle mode of failure (Figure 1 (d)). At higher loads of 900g, detachment of the coating from the substrate within the scratch channel occurred. The appearance of cracks, outside the scratch channel and at an angle of 45 ° ahead of the trailing edge of the stylus, confirmed the more brittle mode of failure of the coating (Figure 1 (e)). At extremely high loads of 2000g, although the depth of the channel was increased, there was no evidence of further damage to the coatings (Figure 1 (f)). This would indicate possible removal and re-deposition of the coating during the course of stylus traverse and/or heavy pressing of the coating itself onto the substrate by the stylus.

At a substrate bias voltage of – 400 V, crack formation both within and outside the channel itself was observed at 200g load, as shown in Figure 1 (g). The number of cracks increased as the load was increased to 900g. Removal of the coating from the substrate outside the scratch channel also occurred at this load, as shown in Figure 1 (h). Figure 1 (i) shows the scratch channel at 2000g load, consisting of regularly spaced cracks within the channel itself at 90° to the direction of stylus travel. These hertzian cracks indicate a brittle mode of coating failure.

At higher substrate bias voltages of –600 V, fine cracks became visible within the scratch channel itself at 400g load (Figure 2 (a)). At higher loads of 1000g, the number of cracks increased and covered the whole width of the scratch channel. Furthermore, the cracks appeared at an angle of approximate 20° ahead of the scratching stylus, as depicted in Figure 2 (b). At 1600g load, spallation of the coating took place outside the scratch channel (Figure 2 (c)). For coatings deposited at both –400 and –600V bias, removal of the coating was possibly due to scratch testing induced failure at highly defective sites.

At –1000V bias voltage, cracking of the coatings followed a different mechanism. At 400g loads, cracks appeared outside the coating and ran almost parallel to the direction of scratching. Although the cracking was not directly related to the channel itself, it can be taken that the cracks were induced as an indirect consequence of the scratching. This is shown in Figure 2 (d). Figure 2 (e) shows the scratch channel at 700g loads. Here cracks were observed at regular intervals along the scratch channel, perpendicular to the direction of stylus travel and outside the scratch channel, at an angle of 45° ahead of the trailing edge of the stylus, typifying the standard failure mechanism observed for these coatings. However no detachment of the coating was observed at any of the applied loads. The formation of such cracks was evident at loads as high as 1900g (Figure 2 (f)).

At –2000 V substrate bias voltage, deformation of the coating was observed at loads as low as 100g (Figure 2 (g)). An increase in the scratching load to 300g and 800g was accompanied by the formation of cracks within the channel and subsequent removal of the coating from the substrate within the channel, as shown in Figure 2 (h) and Figure 2 (i). The mode of failure, showing excessive deformation at 800g load was indicative of a more ductile failure mode, observed for coatings deposited at low bias voltages of 0V and – 200V.

3.2 Variation in Hardness With Substrate Bias Voltage for Nb Coatings

Hardness values for Nb coatings as a function of the bias voltage, including a statistical analysis of five readings for each voltage value (average, standard deviation and maximum/minimum values) are shown in Table 1. At 0V bias, Nb coatings exhibited a hardness of 295 +/– 14 Hv. For coatings produced with the application of a small bias voltage of – 50V, a dramatic rise in hardness (470 +/– 11.5 Hv) was observed. Hardness increased steadily with increasing bias voltage, giving a maximum value of 570 +/– 46Hv at – 200V bias voltage. Further increase in bias voltage was accompanied by a gradual decrease in the coating hardness, as shown in Figure 3, yielding extremely low hardness values, typically 245 +/– 3Hv, at – 2000V bias.

Table 1 *Hardness values of Nb coatings on 316L stainless steel as a function of the bias voltage*

Substrate bias voltage (V)	Microhardness (Hv50)					Average	Stand dev
0	280	310	275	315	296	295.2	14.16
−50	475	462	480	450	485	470.4	11.52
−100	450	432	518	506	487	478.6	30.08
−200	533	491	578	632	616	570.	46.4
−300	597	545	523	564	587	563.2	23.36
−400	522	537	532	560	549	540	11.6
−600	426	475	520	508	463	478.4	28.48
−1000	324	357	360	318	345	340.8	15.84
−2000	243	248	245	251	240	245.4	3.28

4 DISCUSSION

4.1 Coating/Substrate Adhesion

The critical load for removal of the film was generally observed to increase with increasing bias voltage, in the range 0–1000V and the critical load for cracking of the film to occur was a maximum at the high bias voltages of – 600 and – 1000V (Figure 4).

One of the main attributes of the sputtering process is the occurrence of enhanced film-to-substrate adhesion of coatings deposited in such a manner. Such improved adhesion with increasing substrate bias voltage and substrate temperature has been attributed to the formation of a mixed film-to-substrate interfacial region[7] and dense, fine-grained structures[8]. Previous work[9] on the structure of sputtered Nb coatings as a function of substrate bias voltage, revealed that porous, large-grained zone-1 type columnar structures, as reported by Movchan and Demchishin[10] were observed at 0 and – 600V bias, while intermediate bias voltage values of −200 and – 400 V produced finer-grained, dense, zone-1 type structures. Columnar zone-2 and equiaxed zone-3 type structures were observed at high bias voltages of –1000 and

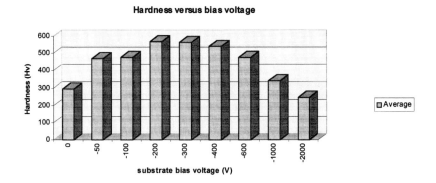

Figure 3 *Average hardness values plotted against substrate bias voltage*

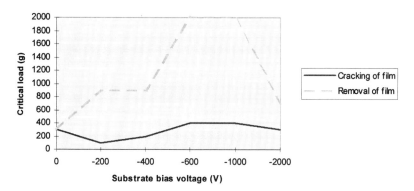

Figure 4 *Variation in the critical load with substrate bias voltage for Nb deposited on 316L stainless steel*

–2000V respectively. In the study here, an increase in the critical load for removal of the coating from the substrate, on the application of a bias voltage (– 200V and – 400V) can be attributed to (i) the formation of a dense, finer-grained structure and (ii) the formation of a graded chemical interface, as a result of enhanced interdiffusion in the film–substrate interfacial region induced by energetic ion bombardment. At –600V, the formation of a larger-grained, porous structure would tend to suggest that the enhanced film-to-substrate adhesion was influenced, to a higher degree, by the formation of a graded interface. This may also be true for coatings deposited at –1000V where no detachment of the film from the substrate was observed, even up to 2000g loads.

4.2 Hardness

The hardness results (Figure 3) showing a marked increase in hardness from 295 to 470Hv, with bias voltage increasing from 0V to –50V, which rose further, attaining a maximum value of 570Hv at – 20,V are similar to those observed for sputter ion-plated TiN coatings deposited on austenitic stainless steel, whereby hardness increased from 900Hv to 2000Hv on the application of a substrate bias voltage of only –10V, which increased to 3000Hv at –50V[11].

The observation on hardness properties of Nb coatings can be interpreted in terms of the microstructure of the coatings, as suggested by Quinto et al[12]. Thus it can be said that in the present study, energetic ion bombardment resulting from intermediate applied substrate bias voltages (–200 V), in conjunction with the high deposition rates (21–23 Å s^{-1}) and low substrate temperatures employed (180–230°C) promoted the formation of finer-grained, highly defected, unequilibriated microstructures. As a consequence, compressive stresses were set up within the coating resulting in higher hardness values.

The increased hardness associated with increasing bias voltage (to –400 V) observed for Nb coatings, may also be attributed to argon gas ions / neutrals incorporated into the coating due to the high gas pressures (1.1 Pa) in conjunction with the intermediate (0–400V) bias voltages employed. Similar mechanisms correlating the hardness with bias voltage/entrapped argon gas have been observed elsewhere[13–15].

The reduction in microhardness in Nb coatings at higher applied bias voltages (– 600 V) was probably due to the formation of larger grained zone-1 structures containing voids, resulting in the transmission of less stress to the neighbouring grains. At – 1000V bias, the formation of large-grained, highly faceted, zone-2 columnar structures probably facilitated the formation of larger indentations due to the large gaps observed between columnar peaks.

Previous work by the authors[16,17] has directly correlated the crystallographic orientations and associated calculated strain energies with the applied substrate bias voltages in the range 0–2000V, for sputtered Nb coatings. Such a relationship is given in Figure 5. It is likely that the formation of {211} crystallographic orientated structures at –200 to –300V bias having higher calculated strain energy than the {222} orientated structures observed at 0V contributed towards the higher hardness values. The formation of zone-1/zone-2 structures at – 600V and –1000V, having {200} crystallographic orientations and therefore having higher calculated strain energies than for {222}/{211} orientated structures would tend to suggest that the lower hardness values were predominantly associated with the morphology of the coating and that the influence of the crystallography of such structures was less marked.

At –2000V bias, further reduction in the observed microhardness can be attributed to thermal recovery due to higher temperatures present at the coating/plasma interface associated with the stored elastic energy within the thermally unequilibriated PVD microstructure. This can be further associated with the onset of a {222} preferentially orientated structure at such bias voltages, having a low calculated strain energy. Additionally, the microhardness of these coatings was less than the values obtained for 316L stainless steel alone. Here, some degree of influence from the potential softening of the substrate occurring at these high bias voltages cannot be ruled out.

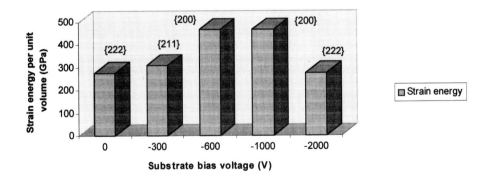

Figure 5 *Calculated strain energy per unit volume as a function of the substrate bias voltage for Nb deposited on 316L stainless steel having {222}, {211} and {200} preferred orientations*

4.3 Correlation Between Hardness and Adhesion

Analysis of the failure modes observed during scratch testing of Nb films deposited on 316L stainless steel showed a systematic correlation with the hardness of the coating. The films deposited at 0V substrate bias voltage exhibited a highly ductile failure mode consisting of material being extruded from the scratch channel and being re-deposited at regular intervals along the channel (Figure 1). This was consistent with the observed low microhardness of these films. For the films deposited at higher substrate bias voltages (–200 V), the observed increase in the microhardness was accompanied by the tendency to undergo failure in a more brittle form, with cracking observed at regular intervals along the scratch channel length. Here, cracking occurred at loads lower than those required to remove the film, in contrast to the coatings deposited at 0V where no cracking was observed prior to film removal. This transformation to a more brittle mode of failure became more evident as the bias voltage increased to –400V and –600V. Although a reduction in hardness was observed as the bias voltage was increased to –400 and –600V, their values were still higher than those obtained for the coatings deposited at 0V. At –1000 V bias voltage however, this typical failure mode was only observed at high loads. At low loads, the formation of cracks outside the channel was observed. Coatings deposited at –2000V showed a sudden transformation to a typical ductile mode of failure, resulting in lower critical loads required for both cracking and removal of the coating respectively. This was reflected in a marked decrease in coating hardness.

Further analysis of the results for Nb indicated a direct correlation between the crystallographic orientation of the coating and the critical load for film removal. Coatings deposited at 0V bias, showing a strong {222} preferred crystallographic orientation, required low loads for their removal. As the bias voltage was increased between 0 and –1000 V, the increase in critical load with increasing bias voltage was accompanied by a transformation from a {222} orientation to a mixed {222}/{211}, through to a predominant {211} orientation, then a mixed {211}/{200} and finally a {200} orientation. However the sudden drop in Lc at –2000V could be attributed to the formation of a {222} orientation. It can be concluded that coatings having preferred crystallographic orientations which tended to favour higher calculated strain energies, required higher critical loads for removal. Clearly, the hardness, deposition process parameters, crystallographic orientation and film adhesion are very much dependent on one another.

5 CONCLUSIONS

1. The hardness of Nb coatings was largely influenced by the substrate bias voltages and the resultant morphology. The application of a bias voltage substantially increased the hardness, which reached a maximum of 560 Hv at – 200 V. This was attributed to the formation a dense, fine-grained, zone-1 structure. The decrease in hardness at higher substrate bias voltages was attributed to the formation of porous, large-grained, zone-1 and zone-2 structures and annealed zone-3 structures.
2. Film-to-substrate adhesion of Nb coatings, measured using the scratch test, was influenced strongly by the substrate bias voltage. The critical load was observed to increase with increasing substrate bias voltage, except for zone-3 type coating structures deposited at – 2000V. However, the low hardness values of the coatings observed at values greater than –600V limited the usefulness of the coating. Enhanced adhesion was attributed to the

formation of a graded interface. Less load was required to promote cohesive failure than adhesive failure, with the exception of films deposited at 0V bias, where the coating was removed without any prior cracking.

3. Direct correlations were observed between the hardness of the coatings and mechanisms of film failure using the scratch test method. Coatings exhibiting higher hardness values showed brittle types of failure, whereas softer coatings failed in a more ductile manner, showing excessive plastic deformation.

References

1. L. P. Ward, PhD Thesis, University of Northumbria at Newcastle, 1991.
2. P.K. Datta, K.N. Strafford and L.P. Ward in 'Corrosion – a Tax forever', Conf. Proc. 28th Australasian Corrosion Association Inc, Nov. 1988.
3. P. K. Datta, K. N. Strafford, L. P. Ward and C. Dick, in . Conf. Proc. I. Mech. E, 14–15 April, London, 1989, p. 179.
4. P. K. Datta, K. N. Strafford, L. P. Ward, in 'Advanced Materials Engineering', D.M.R. Taplin, D. Taylor (eds) IMF 6, Proc. 6th Irish Materials Forum, Dublin, Ireland Sept 6, 1989, Parsons Press, Ireland, 1989, p. 348
5. L. P. Ward and P. K. Datta, In Proceedings Irish Materials Forum 7 conference, Limerick, September, 1991.
6. A .J. Perry, *Surface Engineering*, 1986, **2**, 185.
7. J. E. Greene and M. Pestes, *Thin Solid Films* 1976, **37**, 373.
8. E. Erturk and H. J. Envel, *Thin Solid Films*, 1987, **153**, 135.
9. P. K. Datta, L. P. Ward, R. Hill and K. N. Strafford in 'Surface Engineering Practice – Processes Fundamentals and Applications in Corrosion and Wear', K. N. Strafford, P. K. Datta, J. S. Gray (eds) Ellis Horwood, 1990, p. 89.
10. B. A. Movchan and A. V. Demchishin, *Phys. Metal. Metallogr.*, 1969, **28**, 83.
11. P. J. Burnett and D. S. Rickerby, *Surface Engineering*, 1987, **3**, 69.
12. D.T. Quinto, G. J. Wolfe and P. C. Jindal, *Thin Solid Films*, 1987, **153**, 19.
13. P. Bosland, J. Danroc, R. Gillet and L. Lambard, *Thin Solid Films*, 1988, **166**, 309.
14. T. Lin, K. Y. Ahn, M. C. Harper, P. B. Madaksou and P. M. Fryer, *Thin Solid Films*, 1987, **154**, 81.
15. H. F. Winters and E. Kay, *J. Appl. Phys.*, 1967, **38**, 3928.
16. P. K. Datta, K. N. Strafford, D. S. Lin, L. P. Ward, R. Hill, G. J. Russell, Thin Solid Films, 1988, **168**.
17. L. P. Ward and P.K. Datta, Thin Solid Films (accepted for publication), 1996.

Section 1.4 Fatigue and other Failure

1.4.1
Rolling Contact Fatigue Performance of HVOF Coated Elements

R. Ahmed and M. Hadfield

DEPARTMENT OF MECHANICAL ENGINEERING, BRUNEL UNIVERSITY, UXBRIDGE, MIDDLESEX, UB8 3PH

1 INTRODUCTION

1.1 Background

The practical advantage of high particle speed and temperature offered by advanced thermal spraying processes like Detonation Gun (D-Gun), High Velocity Oxy Fuel (HVOF) and High Velocity Plasma Spraying (HVPS), etc., have enabled them to become an integral part of the automobile and aircraft industry. Some of the applications of these coatings involve their extensive use for clearance control, thermal barrier, and resistance to sliding wear etc. especially in hostile environments. Among these coatings thermally sprayed tungsten carbide-cobalt (WC-Co) coatings deposited by D-Gun and HVOF processes are said to provide a high wear resistance. The performance of these coatings is application dependent and relies on the overall behaviour of the coating and the substrate combination. The advantages of high deposition rates, low cost and the versatility of coating almost any material on almost any substrate call for the investigations of new coating applications. This can enable the industry to exploit the full potential offered by these advanced coating processes. Thermal spray coatings have a lamella structure and are anisotropic in behaviour[1]. These characteristics make analytical investigations like modelling, to evaluate the performance and behaviour of these coatings, very complex and expensive to solve. An experimental approach can thus enable us to investigate the performance and identity of the failure modes of these coatings, for different applications. One such application can be rolling element bearings.

This study addresses the Rolling Contact Fatigue (RCF) performance of thermal spray coatings under various tribological conditions of Hertz stress and lubricant. During this study a modified four ball machine was used to study the performance and failure modes of HVOF coated rolling elements in the geometrical shape of a cone. Three different coating thicknesses were investigated and the results discussed with the aid of Scanning Electron Microscopy (SEM) observations of the failed coatings, microhardness results and stress measurements.

1.2 Previous Studies

Initial studies on the RCF behaviour of thermally sprayed ceramic and metallic coatings were reported by Tobe et al.[2] using a two roller type rolling fatigue test machine. They observed that the surface roughness of the rollers was significant to the fatigue performance, and a roller design to support the coatings at the edges can improve fatigue life. Later studies by Tobe et al.[3] on aluminium substrate revealed that the compressive strength of the coating and the shear stress between the coating and the substrate are the most important factors for the RCF performance of these coatings. The test results showed that the variation in tribological conditions can alter the failure mechanism of these coatings; moreover, significant residual stresses can be generated within the coating microstructure during the RCF tests.

Makela et al.[4] during their experimental approach to study RCF behaviour of WC–12%Co coated steel rollers using a three roller and two roller type test machine revealed that the behaviour of these coatings not only depends upon Hertz contact stress but also on the tribological conditions and vibration during the test. Similarly, Sahoo et al.[5] studied the RCF resistance and WC-Co coatings in the slat and flap track of the aircraft wing under high loads and reported that the cracks initiated from the surface and progressed through the coating thickness. Kershram et al.[6] in their study of the full scale testing of the D-Gun deposited WC-Co coatings for radial bearing sleeves, reported that the coatings were failing in delamination.

In general, these studies show that the coatings fail in delamination and are influenced by

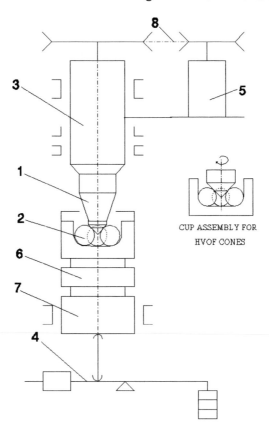

Figure 1 *Modified four ball machine*

1) Coated cone and collet, 2) Planetary balls, 3) spindle, 4) Loading lever, 5) Driving motor, 6) Heater plate, 7) Loding piston, 8) Belt drive

the tribological conditions during the test. However, the studies give a very short account of the effects of variation in coating properties, e.g. coating thickness, and the influence of substrate hardness etc. on the behaviour and performance of these coatings.

2 EXPERIMENTAL PROCEDURE

2.1 Test Configuration

A modified four ball machine as shown in Figure 1 was used to examine the RCF performance of HVOF coated rolling element cones. This machine simulates the configuration of a deep groove rolling element ball bearing. It was used as an accelerated method for the evaluation of fatigue performance as many more stress cycles can be achieved in a fixed contact area on the drive cone. The drive cone represents the inner race and the three planetary balls act as the rolling elements in contact model of the modified four ball machine.

The cup assembly was loaded via a piston below the steel cup from a lever arm load. The coated cone was assembled to the drive shaft via a collet. The cone drove the three planetary balls and the assembly was immersed in the lubricant. A heater beneath the cup can be used to control the lubricating oil bulk at elevated temperatures. The spindle speed may be varied up to 20000 rpm from high or low speed drives. The machine may be set to stop either at revolution number or a maximum vibration amplitude.

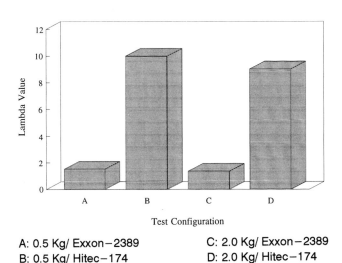

A: 0.5 Kg/ Exxon−2389
B: 0.5 Kg/ Hitec−174

C: 2.0 Kg/ Exxon−2389
D: 2.0 Kg/ Hitec−174

Figure 2 *Elasto-hydrodynamic lubrication results*

Table 1 *Rolling contact fatigue test results*

Test	Hertz stress (GPa)	Lubricant	Average coating thickness (μm)	Time to failure (Minutes)
A	1.72	Hitec-174	150	288*
B	1.72	Exxon2389	150	2188
C	3.08	Hitec-174	150	16
D	3.08	Exxon2389	150	14
E	1.72	Hitec-174	50	3420
F	1.72	Exxon2389	50	1740
G	3.08	Hitec-174	50	14
H	3.08	Exxon2389	50	16
I	1.72	Hitec-174	20	7260
J	1.72	Exxon2389	20	912

* Suspended test

The cup assembly for the test configuration was of type II[7] having a hardness of 60 HRC, whereas the steel planetary balls were grade 10 (ISO 3290-1975) carbon chromium steel with a hardness of 64 HRC and average surface roughness (Ra) of 0.02μm.

2.2 Coated Cone Test Elements

The drive cone tests elements were thermally sprayed by HVOF process to form tungsten carbide cobalt (WC-12%Co) coatings on the surface of mild steel cones having an apex angle of 109.2°. The selection of HVOF coating process for this study was based on the high velocity of the impact particles during the spraying process. The coating material had the advantage of high wear resistance. The rolling elements were thermally sprayed in three different coating thicknesses ground and polished to give an average coating thickness of 150μm, 50μm and 20μm. The substrate cone was sand blasted prior to the coating process to provide a good adhesion by the mechanical interlock at the interface. The average microhardness of the coating material was measured to be 1318 Hv, whereas the average substrate microhardness of the steel substrate was measured to be 218 Hv.

2.3 Test Conditions

Two lubricants were considered for the testing programme. B. P. Hitec-174 is a high viscosity paraffin hydrocarbon lubricant which has a kinematic viscosity of 200 cst at 40°C and 40 cst at 100°C. This lubricant is not commercially available. Exxon 2389 is a synthetic lubricant which has a kinematic viscosity of 12.5 cst at 40°C and 3.2 cst at 100°C. RCF tests were conducted at a spindle speed of 4000 rpm at two different loads of 160N and 400N applied to the cup assembly. Figure 2 represents the Elasto-Hydrodynamic Lubrication (EHL) film thickness results[8] in terms of the ratio of the fluid film thickness to the average surface roughness (λ) for the given test conditions.

3 RESULTS AND SURFACE OBSERVATIONS

3.1 Test Results

Table 1 represents the RCF test results for the rolling element cones in the various tribological conditions. Results are presented in terms of the coating thickness, lubricant type, Hertz contact stress of the uncoated test conditions, and the time to failure. These test results give an appreciation of the performance of the coated elements and are not intended to be used as a basis for the statistical fatigue life prediction. Test results indicate the influence of the Hertz contact stress, coating thickness and lubricant type on the RCF performance of these coatings.

3.2 Surface Observations

Figure 3 shows the SEM observations of the failed WC-Co coated cones for the test D (150 μm thick coating). The figure shows the different stages of the crack propagation leading

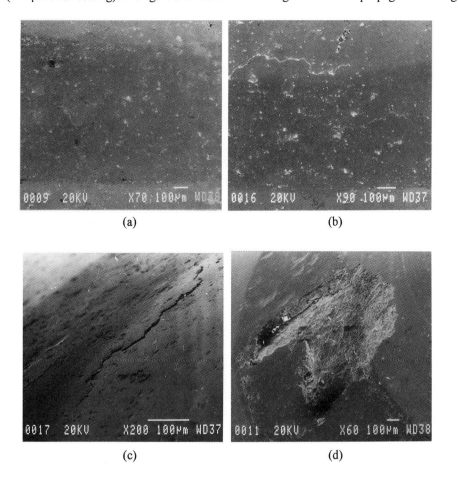

(a)   (b)

(c)   (d)

Figure 3 *SEM test observations of test D*

to the coating delamination. Figure 3 (a) shows the crack initiation within the wear track of the test cone. Figure 3 (b) shows the crack at an advanced stage within the wear track and an appreciation of the occurrence of the edge cracks. Figure 3 (c) shows an inclined view of the cracks shown in Figure 3 (b) and represents the initial stage of coating delamination at the edge of the wear track. Figure 3 (d) shows the failed surface after the coating delamination. Further analysis of the failed coating revealed that the depth of the delamination failure in the middle of the wear track was approximately 80 μm and the Electron Probe Microscopy (EPMA) analysis of the failed coating showed that the coating is failing from within the coating microstructure and not at the interface.

Figure 4 shows the SEM observations of the 50 μm coated cones. Figure 4 (a) shows the Back-scattered Electron Image (BEI) of the failed cone subjected to test G. The figure shows that the coating failure is within the wear track with no edge cracks. The figure also gives an appreciation of the three stages of the coating delamination from the left to the right of the figure. The BEI of the failed coating shows that the coating initially fails just above the interface and the EPMA analysis confirms this behaviour. The figure shows that the coating failure is

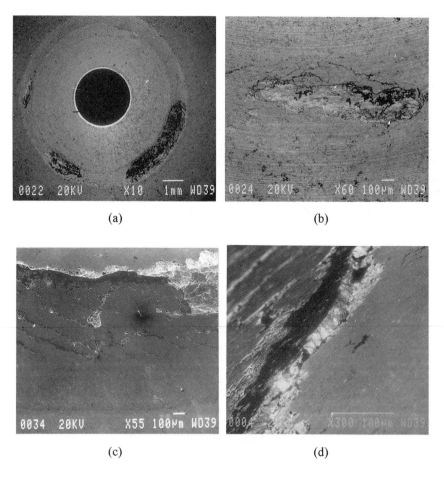

(a)   (b)

(c)   (d)

Figure 4 *SEM observations of test G and H*

initially just above the interface and it is possible that the debris produced during the RCF test caused a three body wear, to wear away the intact coating as the test progresses. This can be seen by the dark and bright portions of the delaminated areas shown in Figure 4 (a). Figure 4 (b) shows the BEI of the initial delamination of the failed coating. Figure 4 (c) shows the failed coating for the cone subjected to test H. The figure shows that the cracks are initiated in the middle of the wear track but, the coating does not fail at the middle of the wear track. It is appreciated that the plastic flow of the substrate from underneath the surface to the edges of the wear tracks takes place due to the ductile behaviour of the mild steel substrate under high compressive stresses. The coating failure takes place at the edge of the wear track. This behaviour is more evident in Figure 4 (d), which shows an inclined BEI of the delaminated edge. The EPMA analysis confirms that the migration of the steel substrate takes place at the edges of the wear track.

Figure 5 represents the detail of cracks shown in Figure 4 (c). It can be clearly seen that the cracks travel around the boundaries of the WC particles. The plastic deformation of the substrate material is thought to be a major factor for the initiation of these cracks in the middle of the wear track.

3.3 Residual Stress Measurements

Rigaku RINT 2000 equipment was used to measure the residual stress fields in the coated rolling elements using the X-ray diffraction $\sin^2\lambda$ method. The technique was used because of its non-destructive nature and the ability to measure macro and micro residual stress on the samples of complex shapes. Residual stress measurements were made on the test cones before and after RCF tests. The measurement results showed an average compressive residual stress of 200 MPa on the surface of pretest cones. The measurements on the wear track of the RCF tested cones showed that the tensile residual stress of the magnitude as high as 100 MPa is generated during the RCF test and magnitude of these stresses varies with the direction of residual stress measurement. The residual stress values are higher when measured along the wear track and their magnitude decreases when measured along the wear

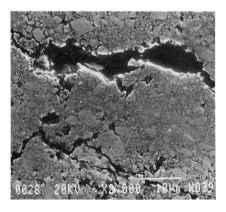

Figure 5 *Crack detail from Figure 4 (c)*

track. Similarly, the residual stress measurements on the pretested cones varies from the cone apex to the base. The behaviour was not observed when similar studies were made on a sample of flat substrate. This can be attributed to the complex substrate geometry of the cones which causes variation in the cooling rates of the coating and the substrate at different sections during the thermal spraying process.

4 DISCUSSION

The RCF test results shown in Table 1 reveal that the Hertz contact stress, coating thickness and lubricant influence the performance of these coatings. Thinner coatings perform better than thick coatings especially at low contact stresses in the presence of high viscosity lubricant. However, the incapability of soft substrate to support the coating at higher stresses leads to their premature failure. The studies on the surface observations of the failed coatings show that the failure mechanism of these coatings varies with the changes in the tribological conditions for the RCF tests. In general the failure of these coatings can be classified into three categories:

Failure type-1   The failure in which the coating delaminates only at the edge of the wear track.
Failure type-2   The failure in which the coating delaminates only in the middle of the wear track.
Failure type-3   The failure of coatings in which the coating delaminates at the edges and in the middle of the wear tracks.

Failure type-1 is rare in the case of coatings on the substrate. However, in some cases this type of failure can be seen on the failed coatings for the samples having coating thicknesses greater than 50 μm. This type of failure is generally seen in the coatings on hard substrates as shown by Hadfield et al[9], and is mainly caused by the tensile stress at the edge of the wear track. Failure type-2 is commonly observed on soft substrate coatings. Figure 4 (a) shows the SEM image of one such failure. This failure type is generally as a result of the plastic deformation of the substrate and the coating cracks in compression due to the lack of support from the substrate. Failure type-3 is generally observed in the case of thick coatings and the cracks originate in the middle and edges of the wear track. Figure 3 (d) gives an SEM image of one such failure. This type of failure can be a result of the combination of two factors, i.e., the ductile behaviour of the coating and the substrate due to the Hertz stress, and the low tensile strength of the coating causing the edge cracks.

Another observation from the SEM studies of the failed coatings is that the thick coatings fail from within the coating microstructure whereas thin coatings fail just above the interface. This is consistent with the previous studies by Ahmed et al[10], in which the WC-Co coatings were failing above the interface and the depth of failure was shown to correspond to the depth of shear stress. Another reason that these coatings do not generally fail at the interface is the good adhesive strength of WC-Co coatings on steel substrate which can be due to the mechanical interlock and, the generation of compressive residual stress at the interface. It also represents a good match between the coating and the substrate properties.

The generation of the residual stress during the RCF tests can be a result of the ductile behaviour of the coating owing to the presence of microcracks and the porosity in the coating microstructure and the shakedown effect caused by the rolling process during the RCF test.

The talysurf analysis of the wear track also shows the ductile behaviour of the coating owing to waviness at the edge of the wear track. In some cases the coating can rise as high as 3μm above the original surface at the edge of the wear track.

In general the tests with the high viscosity lubricant last longer and can be a result of a fully developed EHL film prevailing during the test. Morover, the failures can be affected by the deep penetration of low viscosity lubricant into microcracks and generate a hydrostatic pressure due to the fluid entrapment during the rolling action. However, no studies were done to confirm this mechanism.

5 CONCLUSIONS

1. The failure mechanism of WC-Co coatings during the RCF test is dependent upon the tribological conditions, coating thickness and substrate hardness.
2. Residual stress of tensile nature can be developed during the RCF tests.
3. Thin coatings perform better in low contact stress conditions.
4. The coatings delaminate either from within their microstructure or just above the interface.

Acknowledgements

The authors of this paper acknowledge the support of Professor Shogo Tobe of Ashikaga, Institute of Technology, Japan, for his help in the preparation of coatings. Authors also acknowledge the financial support of Overseas Research Scholarship Scheme which is partly funding the research project.

References

1.   H. Nakahira, K. Tani, K. Miyajima and Y. Harda, Conference Proceedeings, International Thermal Spray Conference, Orlando, Florida, USA 1992, p. 1011.
2.   S. Tobe, S. Kodama and K. Sekiguchi, Conference Proceedings, Surface Engineering International Conference, Tokyo, Japan, 1988, p. 35.
3.   S. Tobe, S. Kodama and H. Misawa, Conference Proceedings, Thermal Spray and Research Applications, 1991, p. 171.
4.   A. Makela, P. Vuoristo, M. Lahdensuo, K. Niemi and T. Mantyla, Conference Proceedings, 'Thermal Spray Industrial Applications', 1994, p. 759.
5.   P. Sahoo, *Powder Metallurgy International*, 1993, **25**, 73.
6.   M. K. Keshran and K. T. Kembayian , Conference Proceedings, National Thermal Spray Conference, Anaheim, CA, USA, 1993, p. 635.
7.   R. Tourret and E. P. Wright, 'Rolling Contact Fatigue Performance Testing of Lubricants' Int. Symp. I. petrolium, October 1976, London, Heyden and Son Ltd, 1977.
8.   B. O. Jacobson, Rheology and Elastohydrodynamic Lubrication, ISBN 0444881468, 1991.
9.   Hadfield M, Ahmed R and Tobe S, Conference Proceedings, International Thermal Spray Conference, Kobe, Japan, May, 1995, p. 1097.
10. R. Ahmed and M. Hadfield, *Surface and Coatings Technology*, in print, 1996.

1.4.2

Fatigue Failure and Crack Growth Response of an Electroless Ni-B Coated Low Carbon Steel

P. B. Bedingfield[1], A. L. Dowson[2] and P. K. Datta[1]

[1]SURFACE ENGINEERING RESEARCH GROUP, SCHOOL OF ENGINEERING, UNIVERSITY OF NORTHUMBRIA AT NEWCASTLE, NEWCASTLE UPON TYNE, UK

[2]IRC IN MATERIALS FOR HIGH PERFORMANCE APPLICATIONS, THE UNIVERSITY OF BIRMINGHAM, EDGBASTON, BIRMINGHAM, UK

1 INTRODUCTION

The process of electroless deposition offers one of the most convenient methods of producing in-situ alloy coatings and has been widely used to produce Ni-P and Ni-B electroless coatings. Alloying arises through the reduction of Ni salts by hypophosphite or boron containing reductants which allow incorporation of P or B in the coating. Such alloying allows the possibility of producing amorphous/microcrystalline coatings with attendant advantages stemming from the absence of grain boundaries and deformation paths. Both the Ni-P and Ni-B coating systems offer these and many other advantages. Indeed the Ni-P coating system is being widely used in many engineering applications to prevent corrosion and wear[1] and represents a widely studied coating type. In contrast, the Ni-B system has not been studied extensively and has not been accepted for application in spite of its many potential advantages. Such advantages are reported to include higher hardness and wear resistance[2-4], higher melting points, and improved solderability[6-8]. In many circumstances, however, the benefits offered by a coating layer in providing wear and corrosion resistance are offset by a corresponding deterioration in fatigue properties. Coating residual stresses, undesirable microstructures and defect structures – and in particular the facility of interfaces as potential crack nucleation sites – can all lead to lower fatigue life and crack growth resistance. Despite these observations, and the growing use of coatings in both corrosion and fatigue critical applications, fundamental studies relating to the effects of coating layers on fatigue performance have been limited and in particular, there is little information relating to the initiation and growth of fatigue cracks in coating layers.

The current paper details some recent observations relating to the fatigue performance of 30 μm thick dimethylamineborane (DMAB) reduced electroless Ni-B (3–5 wt % B) coatings deposited on a mild steel substrate and reviews some provisional experimental observations relating to the initiation of fatigue cracks within the coating layer and their subsequent growth across the coating/substrate interface. Particular attention is given to the effects of post deposition heat treatments and the results are rationalised in terms of the observed variations in microstructural morphology and the associated changes in residual stress distribution within the coating layer.

2 EXPERIMENTAL

2.1 Materials Preparation and Coating Deposition

With the exception of the in-situ test specimens used in the thin strip deflection residual stress studies, all specimens used in the coating deposition and fatigue studies were fabricated from a 0.15% C, 0.80% Mn steel (analytical specification BS970 – 080A15). Specimen surfaces were initially pre-ground to a uniform surface finish using 240 grade SiC paper, soaked in a commercial cleaning agent (Leaprep 20) at 70°C for 3–5 minutes, and then anodically cleaned (in the same solution) at 70°C for 1-2 minutes. After cleaning each specimen was rinsed in water, etched for 1–2 minutes in 30% Hydrochloric acid and then subjected to a further rinse before transferring to the plating bath. Coatings were deposited from a laboratory bath containing a proprietary solution (Niklad 740) operated at pH6/72°C – conditions which had been determined as optimum from earlier studies of resulting surface morphologies and quantitative measurement of coating - substrate adhesion[9]. Deposition kinetics under these conditions had been determined previously and allowed accurate control of coating thickness with deposition time. After deposition, selected specimens were subjected to one hour post-deposition heat treatments at 300, 450 and 600°C in order to modify the coating microstructure and vary the level and distribution of residual stress across the coating/substrate interface. The effects of such heat treatments on the structure and morphology of the coating layer have been discussed in detail elsewhere and will only be considered briefly in the current paper. All coatings, whether as deposited or heat treated, were examined using standard optical and SEM procedures and the extent of crystallisation and structural development were assessed using XRD analysis.

2.2 Fatigue Life and Crack Growth Studies

The effects of as-deposited and heat treated coatings on the resistance of the substrate material to failure under cyclic loading conditions were assessed using an Avery duplex rotating bend fatigue test machine operated in air at a frequency of 3000 rpm. Test specimens were subjected to constant amplitude loading conditions ($R = \sigma_{min}/\sigma_{max} = -1.0$) with peak stress ($\sigma_B$) values ranging from approximately 275 up to 400 MPa (~70–100% of the uniaxial yield stress of the substrate material). In all instances the fatigue limit was assumed to have been reached if the total number of cycles required to cause failure exceeded 10^7 cycles. In addition to these studies the fatigue crack initiation and crack growth response of the as-coated material were examined using a Dartec servo-hydraulic testing machine with test specimens being subjected to constant amplitude, three-point bend loading ($R \sim 0.1$) in air at a frequency of 25 Hz, with peak stress values being selected in order to induce failure within 10^5 cycles. The crack initiation response of the material and the subsequent growth of fatigue cracks within the coating layer were monitored intermittently during the test using the acetate replication technique with measurements being taken at mean load and at approximately 1000–5000 cycle intervals.

2.3 Residual Stress Determination

The extent of residual stress build-up in the coating layer was determined using either X-Ray diffraction or 'thin strip deflection' techniques. As pointed out by Celis-et-al[10] the amorphous and/or microcrystalline nature of electroless Ni-B deposits introduce major problems in terms of residual stress determination because the X-ray diffraction patterns which are generated are very often diffuse and normally suffer from extensive line broadening. Consequently XRD techniques based on line broadening and peak displacement measurements are intrinsically inaccurate and can only be applied where the crystal structure is fully developed (i.e. after heat treatment) or where individual diffraction peaks can be accurately identified. For this reason the extent of residual stress build-up in the as-deposited coatings was estimated on the basis of in-situ thin-strip deflection measurements which were made during the deposition process. The technique used in the current work is based on the method developed by Dvorak and Vrobel and assumes that residual stresses which build up in the deposit will induce an expansion or contraction in a thin strip substrate of known dimensions which can be equated with the magnitude and sign of the induced load. Strip displacement was monitored at approximately 15 minute intervals during deposition via a dilatometer and all clamps, loading arms and support rods were coated with a protective PTFE coating to ensure that deposition occurred on the specimen only. To ensure accuracy of measurement each test was carried out in duplicate using a freshly prepared plating bath and different strip lengths were assessed in order to test the validity of the technique.

The levels of residual stress within the coating layers after heat treatment were assessed using the Psi Diffraction technique[12,13]. This technique is based on the fact that there is a direct correlation between the 'd' spacing of the atomic lattice and the level of internal stress. As such the sign and magnitude of the residual stress can be determined from angular shifts in the position of a pre-selected X-ray diffraction peak which in turn can be related back to a specific crystallographic plane. All measurements were conducted on a Siemens D500 X-ray Diffractometer and the X-ray diffraction profiles obtained were evaluated using the Siemens "Stress AT" software. Cr Kα radiation ($\lambda = 2.290$ Å) was used throughout the experimental work enabling measurements to be obtained for the Ni (220) peak at a 2θ value of ~ 133°. The use of such high 2θ values increases the sensitivity of the Psi Diffraction technique since the angular shift for relatively small changes in 'd' becomes more pronounced the higher the 2θ value. Psi offsets (angular position of the diffracted beam with respect to the incident beam) ranging from -42.13 up to $+45.00$° were employed, producing 11 separate measurements for each specimen examined, and diffraction profiles were acquired between 130 and 136° 2θ using a step size of 0.1° and an acquisition time of 8 seconds per step.

3 RESULTS

3.1 Coating Layer Morphology

The morphological and topographic features of the coating layers after deposition and the effects of subsequent post-deposition heat treatments on microstructural development have been discussed in detail elsewhere[9]. Summarising these observations, the deposition process was characterised by the development of isolated nuclei on the specimen surface followed by lateral growth and coalescence to form a continuous but somewhat uneven surface film. Typical SEM micrographs illustrating the surface morphology and the through thickness microstructural characteristics of the as-deposited coating layers are reproduced in Figure 1. In all instances a near-amorphous, nodular type growth morphology was observed to develop with occasional columnar-type features extending across the thickness of the coating to the

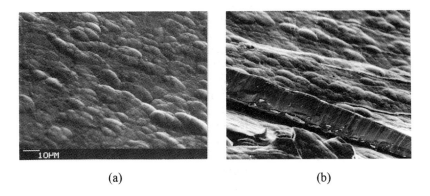

(a)   (b)

Figure 1 *SEM micrographs showing (a) the surface topography and (b) the through-thickness microstructural characteristics of the as-deposited coating layers*

coating/substrate interface. In keeping with these observations X-ray diffraction analysis revealed the presence of a micro-crystalline (crystallite size ~ 40–90 Å), supersaturated Ni-based solid solution with a preferred {200} orientation. X-ray diffraction profiles obtained for the as-deposited coating are presented in Figure 2. The principal effect of the post deposition heat treatment was to promote a breakdown in the supersaturated solid solution and in all cases as the time and temperature of exposure were increased there was a progressive reduction in the extent of line broadening, an apparent increase in the crystallite size and an associated strengthening of the Ni {200} texture. As indicated in Figure 3 for coatings heat treated at 450°C such changes were accompanied at the higher temperatures of exposure by the nucleation and growth of Ni_3B precipitates and morphologically, by a gradual smoothing of the coating surface topography. The ageing response of the coating layers and the associated changes in crystallite size and microhardness have been discussed in some detail elsewhere[9].

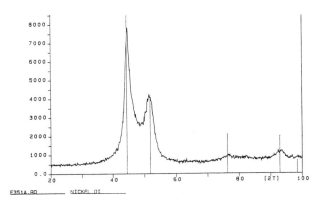

Figure 2 *X-ray diffraction pattern obtained from the as-deposited coating showing extensive line broadening and preferred Ni{200} orientation*

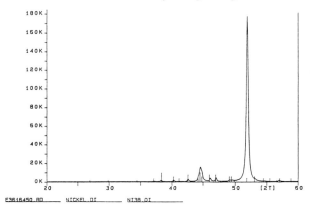

Figure 3 *X-ray diffraction pattern obtained from the coating layer following heat treatment at 450°C for 16 minutes showing reduced line broadening and indicating the clear presence of Ni₃B precipitates*

3.2 Fatigue Life and Crack Growth Behaviour

Fatigue life ('S' Vs 'N') data obtained for the substrate material and coated specimens tested in the as-deposited condition are compared in Figure 4. It is apparent from these results that the coating layer had a marked effect on the fatigue response of the material with the 'S' Vs 'N' curve being shifted both vertically and horizontally indicating an effective reduction in both the fatigue limit (minimum stress required to cause failure) and the number of cycles (N_f) required to cause failure. The effects of ageing treatments on the fatigue life of the coated material are demonstrated in Figure 5. A progressive recovery in the fatigue performance of the material was observed following heat treatment with the fatigue ('S' Vs 'N') curves obtained from coated materials heat treated at 450 and 650°C becoming almost coincident

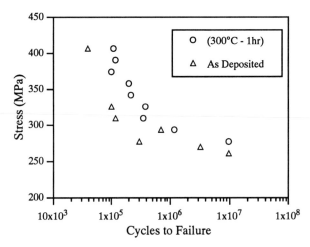

Figure 4 *'S' Vs 'N' data showing the effect of a 30 µm thick electroless Ni-B (3–5% B) coating on the fatigue response of a low carbon (0.15%C/0.80%) mild steel*

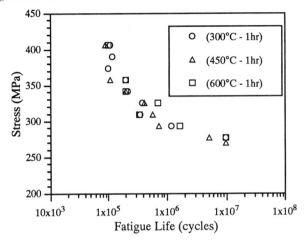

Figure 5 *'S' Vs 'N'data showing apparent recovery in fatigue behaviour for electroless Ni–B coated specimens following heat treatment at 450 and 600ºC respectively*

with that obtained for the untreated substrate material. Results obtained for coatings heat treated at 300ºC were inconclusive and appeared to indicate an intermediate fatigue response with fatigue lives lying mid-way between those recorded for the untreated substrate material and coatings tested in the as-deposited condition.

The morphological and microstructural features associated with the crack initiation process and the subsequent growth of fatigue cracks across the coating/substrate interface were assessed using acetate replicas taken at approximately 5000 cycle intervals from the surface of planar specimens which had been subjected to constant amplitude three point bend load cycling (R ~ 0.1). Representative micrographs obtained from these studies are reproduced in Figure 6 for cracks formed in the as-deposited coating layer. Within the initial stages of life load cycling was found to be accompanied by multiple crack initiation, primarily at microstructural inhomogeneities within the coating layer, and by a general increase in crack density. As a general rule, however, the majority of these cracks were found to be non-propagating and,

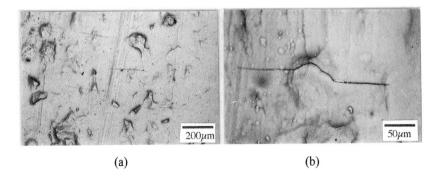

(a)   (b)

Figure 6 *Optical micrographs obtained from three point bend surface replica studies showing (a) multiple fatigue crack initiation and (b) fatigue crack initiating at a surface irregularity in the coating layer*

after an initial rapid burst of growth, they failed to play any further role in specimen failure. Typical crack growth rate, dc/dN, versus half surface crack length, c, plots obtained for three such cracks are reproduced in Figure 7. The main point to note from these data is that all three cracks exhibited a marked deceleration in growth rate as the crack length approached 40 μm and at crack lengths in excess of 50 μm the crack either arrested or the crack growth rate reached such a low value that it became indeterminate. Whilst the aspect ratio (half surface crack length/crack depth (c/a) ratio) was not measured in the current work it is significant to note that, if we assume the aspect ratio to be of the order 0.7–1.0 (a reasonable assumption considering the crack size and the test specimen geometry) the point at which crack retardation and/or arrest occurs appears to be coincident with the point at which the crack front approaches the coating/substrate interface.

The microstructural features associated with crack formation and the early stages of growth for a crack which was observed to have penetrated the coating/substrate interface and which ultimately led to specimen failure are illustrated in Figure 8. In this instance the crack was found to have initiated within the bounds of a large nodular defect on the surface of the deposit and had propagated in a planar manner interacting with neighbouring cracks both ahead of and around the main crack tip. Associated da/dN data obtained for this crack are plotted in Figure 9 as a function of the nominal applied stress intensity range (ΔK_1). ΔK_1 values were calculated on the basis of the projected crack length measured perpendicular to

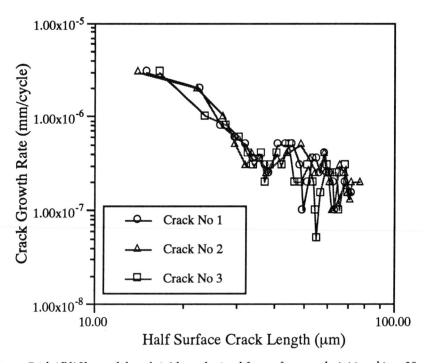

Figure 7 *'dc/dN' Vs crack length 'c' data obtained for surface cracks initiated in a 30 mm thick as-deposited electroless Ni-B coating layer showing crack growth rate retardation and crack arrest with increasing crack length*

the stress axis using solutions derived by Scott and Thorpe[14] such that for the deepest point of the crack:

$$\Delta K_1 = M_{f(\pi/2)}\left[1 - 1.36(a/t)(a/c)^{0.1}\right]\frac{\Delta\sigma_b\sqrt{\pi a}}{E(k)}$$

where:

$M_{f(\pi/2)}$ = The front face correction factor $(1.13 - 0.7(a/c)^{1/2})$
$E(k)$ = The elliptical integral of the second kind $[1 + 1.47(a/c)^{1.64}]^{1/2}$
a = crack depth; c = half surface crack length; t = specimen thickness; $\Delta\sigma_b$ = total stress range.

Correcting for the finite size of specimen, the effective stress intensity at the crack tip is given by:

$$\Delta K_{1eff} = \frac{\Delta K_I\left[1 + F(a/c)G(c/w)H(a/t)\right]}{(0.2745)^2}$$

where w is the specimen width and $F(a/c)$, $G(c/w)$ and $H(a/t)$ are polynomial expressions derived by Holdbrook and Dover[11].

It should be noted that the data presented in Figure 9 were generated under progressively increasing loading conditions (444, 468, and 484 MPa) and consequently it was likely that this loading sequence was instrumental in driving the crack beyond the coating/substrate interface. The results are, nevertheless, consistent with those presented in Figure 6 for nonpropagating cracks in that a distinct barrier to crack propagation was evident, as indicated by the initial deceleration in crack growth rate and by the apparent minimum in the da/dN Vs ΔK_1 curve. As the stress intensity level was increased beyond approximately 7 MPa√m, however,

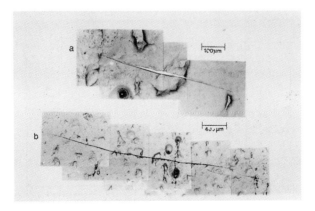

Figure 8 *Optical micrographs showing the microstructural features associated with crack initiation and the early stages of growth for the crack which ultimately led to specimen failure*

Table 1 *Psi diffraction residual stress and surface microhardness data obtained for as-deposited and heat treated coating layers*

Teat treatment	Residual stress (MPa)	Hardness (Hv)
As-deposited	-	807
1Hr – 300°C	+102 ± 60	1042
1Hr – 450°C	–517 ± 29	700
1Hr – 600°C	–387 ± 21	386

crack extension was allowed to occur and a progressive increase in the crack growth rate was observed. The phenomenological and microstructural factors contributing to this initial deceleration and subsequent acceleration in crack growth rate are not directly apparent from the metallographic observations presented in Figures 6 and 8 and in this respect it is clearly difficult to distinguish between a propagating crack and a nonpropagating crack, since in both cases the morphological features associated with crack initiation and the early stages of development are essentially the same. These results would tend to suggest that although crack initiation had occurred within the coating layer, the subsequent extension of the crack across the coating/substrate interface was facilitated by the development of a secondary crack which had originated at a sub-surface inclusion or microstructural discontinuity beneath the coating layer.

3.3 Residual Stress Measurements

The results of residual stress measurements obtained for the as-deposited coating using the in-situ 'thin strip deflection' method indicated relatively high levels of stress (up to 80

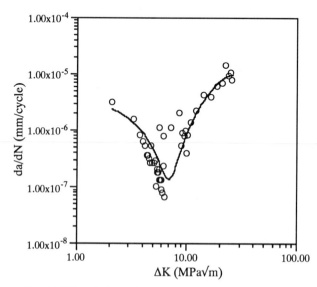

Figure 9 *Typical 'da/dN' Vs ΔK data obtained for a self-initiated fatigue crack which was observed to have penetrated the coating/substrate interface*

MPa) within the coating layer at the start of deposition. However, as the coating layer developed the stress level was observed to decrease significantly falling to below 20 MPa as the coating thickness was increased beyond 8 μm. Good agreement was observed between measurements obtained for different strip lengths and results consistently indicated residual tensile stresses in the coating layer, regardless of the coating conditions (pH level, bath temperature etc.) or the coating thickness. In contrast, X-ray diffraction measurements of the residual stress levels in the heat treated materials indicated a transition from residual tensile to residual compressive stress as the temperature of heat treatment was increased with the residual compressive stress reaching a maximum (−517 ± 29 MPa) in the material heat treated at 450°C. Typical results obtained for 30 μm thick coatings heat treated for 1 hour at 300, 450 and 600°C are reproduced in Table 1 together with surface microhardness values. It is evident from these results that coatings heat treated at 300°C retained a high level of tensile stress (102 MPa). However, these results were subject to a large degree of scatter (± 60 MPa), primarily due to the extensive line broadening observed in the diffraction patterns (Figure 2) obtained for this material and the associated uncertainty regarding the determination of 'd' spacings (and hence calculated strain values). This contrasts the results obtained for coating layers heat treated at 450 and 600°C where, consistent with the observed sharpening of the X-ray diffraction profiles, the scatter was reduced to within 29 MPa. As indicated in Table 1 the apparent change from residual tensile to residual compressive stress with increasing temperature was accompanied by distinct changes in surface microhardness. Heat treatment for 1 hour at 300°C was accompanied by a significant increase in microhardness compared to the as-deposited coating layer with hardness levels exceeding 1000 Hv. As the temperature was increased to 450 and 600°C, however, a considerable reduction in microhardness was observed suggesting significant overageing of the coating layer.

4 DISCUSSION

Electroless Ni deposits (both Ni-P and Ni-B) have been traditionally developed as protective layers for use in a wide range of structural and engineering applications where there is a need for improved corrosion and/or wear resistance. Whilst the corrosion response of these deposits has been extensively studied and the deposition parameters which influence microstructural development and coating performance are well documented, little information is available regarding the parameters which influence the mechanical performance of the coating layer, particularly under conditions where the component may be subjected to mechanical and/or thermal load cycling. It is, nevertheless, generally conceded that coatings developed for corrosion resistance, although affording good protection against environmental degradation, can adversely affect the fatigue performance, particularly at ambient and low temperatures. However, the mechanisms involved have not been clearly identified and there is a need to characterise and quantify the effects of coating layers on the processes of crack initiation and crack propagation. The current studies have highlighted the adverse effects of electroless Ni-B (3–5 wt% B) coatings on the fatigue performance of a low carbon steel and have indicated that for a 30 μm thick coating reductions in fatigue life of up to 50% can be observed (Figure 4). Significantly however, a distinct recovery in fatigue performance was observed following heat treatment and in this respect fatigue lives observed for materials heat treated at 450 and 600°C were found to be almost coincident with those obtained for the untreated substrate material (Figure 5). In general these results would appear to be in good agreement with residual stress measurements obtained for the various coating layers and with the associated microstructural changes and reductions in microhardness which accompanied heat treatment. In-situ thin strip deflection measurements of the residual stress distributions within the coating layer conducted during the deposition process indicated relatively high levels of residual tensile stress which, when coupled with the high hardness (≥ 800 Hv) and apparent low ductility of the coating layer, would have been expected to facilitate both crack initiation and growth within the coating layer. This clearly contrasts the observations made for the heat treated materials where X-ray diffraction measurements and microhardness studies indicated the build up of residual compressive stresses in the coating layer, significant reductions in microhardness (≤ 400Hv at $600°$C) and the simultaneous nucleation and growth of Ni_3B precipitates. Consistent with these observations the metallographic studies conducted on replicas taken from the surface of three point bend specimens coated with a 30 μm thick as-deposited electroless Ni-B coating layer have shown (Figure 6) that fracture is preceded by the development and growth of multiple fatigue cracks which initiate both at microstructural discontinuities within the coating layer and at nodular like defects which extend across the coating to the coating/substrate interface. Failure then occurred by the subsequent growth and interaction of these fatigue cracks and by their subsequent penetration into the underlying metal substrate. As a general rule, however, the majority of these cracks were found to be non-propagating and, after an initial rapid burst of growth, a distinct reduction in fatigue crack growth rate was observed to occur, primarily as a result of the crack tip interacting with the coating / substrate interface (Figure 7). In this respect the interface was found to act as a significant barrier to fatigue crack propagation. Similar observations have been made by Suresh[16] who suggested that the tendency for fatigue cracks to arrest could be rationalised in terms of the interactions between the interface and the crack tip plastic zone and as such could be interpreted in terms of associated changes in crack tip opening displacement. In this respect for a fatigue crack propagating from a cyclically weak to a cyclically strong material, the constraint imposed by the stronger material ahead of the crack tip would both impede slip within the plastic zone and locally reduce the crack tip opening stresses, causing the crack to arrest before penetrating the interface. It should be noted however, that these consideration do not take into account the residual stress distributions within the coating and substrate layers. In the present materials the high levels of residual tensile stress within the coating layer would have been balanced by compressive stresses in the metal substrate which, as the crack approached the coating/substrate interface, would act to reduce the crack tip opening displacement causing a reduction in crack growth rate. Such reductions in crack growth rate are clearly reflected both in the 'dc/dN' Vs 'c' data presented for non-propagating cracks presented in Figure 7 and in the 'da/dN' Vs 'ΔK' curve obtained for the crack which ultimately led to failure. On this basis ultimate fracture appeared to be governed, not specifically by the initiation and growth of short fatigue cracks within the coating layer, but rather by the interaction of these cracks with underlying defects and/or microstructural discontinuities within the metal substrate. Such interactions would have acted to increase defect size thereby increasing the driving force for crack propagation above the threshold for long crack propagation within the substrate material.

5 CONCLUSIONS

1. The effects of a 30 μm thick electroless Ni-B coating layer on the fatigue response of a low carbon (0.15% C, 0.80% Mn) mild steel have been studied, both in the as-deposited

condition and following heat treatment. Whilst reductions in fatigue life of up to 50% were observed for material tested in the as-deposited condition, a progressive recovery in fatigue behaviour was observed following heat treatment at 450 and 600°C with fatigue lives becoming almost coincident with those observed for the untreated substrate material.

2. The apparent reduction in fatigue life observed for materials tested in the as-deposited condition can be attributed, in part, to the high levels of residual tensile stress which were observed to develop in the coating layer during deposition. This contrasts the observations for the heat treated material (450 and 600°C) where XRD measurements indicated a build up of compressive residual stress, possibly due to the precipitation and growth of Ni_3B.

3. The principal factor contributing to the reduced fatigue life in the as-coated material was the increased incidence of crack initiation. As-deposited coating layers were observed to facilitate the crack initiation process resulting in multiple crack formation. Whilst the majority of these cracks were observed to arrest at the coating/substrate interface, ultimate failure was promoted by a combined process involving multiple crack tip interaction and coalescence, and by the interaction of the crack tip stress field with subsurface defects beneath the coating/substrate interface.

4. Crack growth data obtained for surface cracks initiated within the coating layer indicated that these cracks could propagate at ΔK values significantly below the long crack threshold for the substrate material. A distinct retardation in growth rate was, however, observed as the crack tip approached the coating / substrate interface and in the majority of instances cracks failed to penetrate into the underlying substrate.

References

1. S. A. Watson, Nickel Development Institute Technical Report No 10055.
2. K. N. Strafford, P. K. Datta and A. K. O'Donnel, Materials and Design, 1982, **3**, 608.
3. A. K. O'Donnel, M. Phil., (CNAA) Newcastle Upon Tyne Polytechnic 1982.
4. K. M. Gorbunova, M. V. Ivanov and V. P. Moiseev, *J. Electrochem. Soc.*, 1973, **120**, 613.
5. Elliot, 'Constitution of Binary Alloys', McGraw Hill Book Co. Inc, 1965, 127,661.
6. D. W. Baudrand, Proc. 66th Annual Technical Conference of the Am. Electroplaters Society, Atlanta 1987, Paper 2.
7. N. Felstein and T. G. Thomann, *Plating and Surface Finishing*, 1979, **66**, 50.
8. J. Shrantz, *Industrial Finishing*, 1983, **59**, 23.
9. P. B. Bedingfield, D. B. Lewis, P. K. Datta, J. S. Gray and P. B. Wells, *Trans. Inst. Metal Finishing*, 1991, **700**, 19.
10. J. P. Celis, J. R. Roos and M. DeBonte, Proc. Conf, Surface Modification Technologies, Pheonix, Arizona, 1988, p215.
11. A Dvorak and L. Vrobel, *Trans. Inst. Metal Finishing*, 1971, **49**, 153.
12. I. C. Noyan and J. B. Cohen, 'Resudual Stress – Measurement by Diffraction and Interpretation', Springer-Verlag, 1987.
13. B. D. Cullity, "Elements of X-Ray Diffraction", 2nd Ed., Addison Wesley, 1978, 447.
14. P. M. Scott and T. W. Thorp, *Fatigue of Engineering Materials and Structures*, 1981, **4**, 291.
15. S. J. Holdbrook and W. D. Dover, *Engineering Fracture Mechanics*, 1979, **12**, 347.
16. S. Suresh, Y. Sugimura and T. Ogawa, *Scr. Metal et Mater.*, 1993, **29**, 237.

1.4.3

Observations and Analysis of a 200nm Aluminium Film on Silicon Using Ultra Micro-Indentation

A. J. Bushby[1] and M. V. Swain[2]

[1]DEPARTMENT OF MATERIALS, QUEEN MARY AND WESTFIELD COLLEGE (UNIVERSITY OF LONDON), MILE END ROAD, LONDON E1 4NS, UK

[2]CSIRO, DIVISION OF APPLIED PHYSICS, LINDFIELD, NSW 2070, AUSTRALIA

1 INTRODUCTION

Aluminium is widely used in commercial applications as a thin film coating because of its ease of deposition. These coatings find use for the reflective surfaces of optical mirrors, low electrical resistance paths in semiconductors and for interconnections as well as for the coating of polymeric packaging materials used for the preservation of food stuffs in multilayer systems.

Invariably the near-surface behaviour of such coatings is dominated by the presence of a thin oxide layer which is both much harder and more brittle than the underlying substrate. Chechenin et al[1] have recently investigated the behaviour of such layers on aluminium. The range of oxide film thicknesses was 3–360 nm.

The emphasis of this paper is to compare two approaches for the estimation of the mechanical properties of an aluminium film on silicon. In the first use is made of pointed indenters, typically a Berkovich, and in the second approach uses of a small radius (5 mm) spherically tipped indenters. In both cases the hardness versus depth of penetration will be compared. Previous studies have shown that in principle with the use of spherically tipped indenters the ability to determine the developing elastic/plastic behaviour is possible whereas with a pointed indenter the behaviour is that of nearly constant equivalent strain (Tabor[2]). More recently the application of these traditional approaches has been improved by the development of depth sensing indentation machines which can record force/penetration curves during contact event.

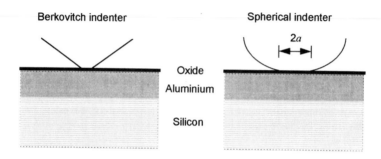

Figure 1 *Schematic of the indenter and material configuration*

2 BACKGROUND

The contact of a pointed or spherically tipped indenter onto a film deposited on a substrate is shown schematically in Figure 1.

The resultant stress field is very complex because of the difference in the elastic properties associated with the film and substrate. Typically finite element numerical techniques (FEM) have been employed to determine the complex deformation and force-displacement behaviour of these systems when loaded with pointed indenters[3,4]. There have been some recent numerical analyses possible for the sphere on a film on a substrate system by Schwanger et al[5,6]. Also Gao et al[7] have provided a relatively simple procedure to determine the influence functions for such thin films on substrates particularly for the determination of modulus effects. A critical review and application of such procedures has been carried out recently by Mencik et al[8] for a wide range of materials on two different substrates. Mencik et al[8] also found that there was often a difference between flexure and indentation determined modulus E, because of texture associated with most film deposition and growth.

A critical feature of the FEM and analytical contact solutions is the two scalar parameters namely the radius of the contact circle, a, and the film thickness, t (see Figure 2). Only for film thickness eight times the contact circle radius ($t \geq 8a$) does one expect a measure of the film properties almost independent of those of the substrate[6]. However, particularly with the situation of stiffer brittle films on softer substrates, film cracking occurs rather than plastic deformation of the film. It is also possible to generate plastic deformation of the substrate while maintaining elastic deformation of the film. This and other possibilities were qualitatively considered in a recent review by Swain and Mencik[9]. The role of both film thickness and substrate on the hardness and modulus of TiN films was considered by Wittling et al[10].

2.1 Pointed Indenters

The Berkovich indenter is loaded in continuous sequence to a maximum load at constant strain

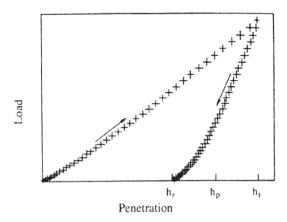

Figure 2a *Continuous load-unload sequence (Berkovich indenter)*

rate followed by a continuous unloading sequence to zero load. The type of force/penetration curve generated is illustrated schematically in Figure 2a.

The analysis of the data is performed in the manner outlined by Oliver and Pharr[11]. The hardness is determined as

$$H = F/A \qquad \{1\}$$

where F is the applied force and A is the projected area of the contact and this is calculated from the depth of penetration below the contact, h_p (Figure 2a),

$$A = 24.5 \, h^2 \qquad \{2\}$$

for a Berkovich or Vickers indenter. The hardness can be calculated as a continuous function of depth by calculating the elastic contribution of the total depth and hence the depth below contact, h_p. The elastic displacement is assumed to be equally divided above and below the contact[11,12]. This can be achieved by assuming the elastic contribution is equivalent to that of a flat punch with the same projected area[11]. Hence hardness as a function of depth of penetration can be found.

2.2 Spherical Indenters

For spherical indenters a novel load-partial unload sequence is used developed by Field and Swain[12] in which after each increment of load the indenter is unloaded to half this value before proceeding to the next increment of load (Figure 2b).

The unloading is assumed to be entirely elastic (Hertzian) and equivalent to the reloading of an elastic preformed cavity having the form of the residual impression[12]. The analysis developed by Field and Swain[12] can be used to calculate the residual displacement, h_r, and hence find the depth below contact, h_p, for each pair of load / unload data points. Again the elastic displacement is

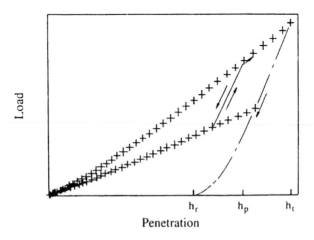

Figure 2b *Load-partial unload sequence (spherical indenters)*

assumed to be equally divided above and below the contact. If the two curves coincide the response is entirely elastic and at the onset of plastic or permanent deformation the curves diverge.

The hardness or contact pressure, P_m, is determined from:

$$P_m = \frac{F}{\pi a^2} \qquad \{3\}$$

where F is the applied force and a is the radius of the circle of contact given by

$$a = \sqrt{2Rh_p + h_p^2} \qquad \{4\}$$

where R is the radius of the indenter. For fully elastic contacts where $h_t \ll R$, Equation $\{4\}$ approximates to

$$a^2 = R\,h_t \qquad \{5\}$$

where h_t is the total penetration depth under load. Substituting Equation $\{5\}$ into $\{3\}$ gives

$$P_m = \frac{F}{\pi R h_t} \qquad \{6\}$$

For a full analysis see reference 12.

3 EXPERIMENTAL DETAILS

The aluminium films were deposited onto a silicon wafer by electron beam evaporation. The thickness of the film deposition was monitored during deposition and found to be 200 nm. After deposition the samples were exposed to normal laboratory atmosphere.

The mechanical properties were determined using a UMIS-2000 commercially available ultra-micro indentation system. Details of the system have been published elsewhere[13]. This system readily enables changes of indenter tips and in this study both a Berkovich and spherically tipped (5mm radius) indenters were used. The tips were characterised to determine the area of contact versus penetration depth and the algorithm used to analyse the data together with the relationships given in Section 2. In the case of the spherical tipped indenter, using a procedure developed by Bushby et al[14], the effective radius versus depth of penetration below the circle of contact was determined. For both indenters optical grade fused silica was used for the calibration.

Force-displacement curves where generated using the continuous sequence (Figure 2a) for the pointed indenter and the load-partial unload sequence (Figure 2b) for the spherical indenter. Measurements were made on the silicon with and without the aluminium film using both types of indenter.

4 OBSERVATIONS

4.1 Pointed Indenter

Typical force-displacement curves for the silicon and the aluminium film on the silicon substrate with the Berkovich indenter are shown in Figures 3 (a)-(d). The observations on the silicon are shown in Figure 3(a), in which five sets of results are superimposed and show excellent reproducibility with a deviation of the maximum penetration depth of only 1-2 nm. Figure 2b shows an individual curve. Close inspection of the very low load/shallow displacement region of the indentations into silicon reveals a small discontinuity at a load of 0.2mN and ~15 nm penetration. Note also the distinct change in slope of the unloading curve at ~35% of the maximum load. This issue will be taken up in the discussion.

The force-displacement curves for the pointed indenter loading onto the aluminium film (Figures 3(c) and 3(d)) show a very different type of behaviour. At very low loads the resistance to penetration is very striking but at a load of ~0.2 mN there is a sharp discontinuity with significant penetration (~30nm) before the curves follow a more gradually increasing resistance to further penetration. The onset load for the discontinuity ranges from ~0.13 to 0.25 mN and is thought to be associated with rupture of the oxide film on the aluminium. The results for the indentations on the aluminium show a greater scatter than the impressions on the uncoated silicon, typically (~3 nm) which is associated with the greater roughness of the aluminium film and the variation in the oxide film fracture load. Note also the very steep and more continuous unloading curves in the case of the aluminium. In Figure 3d the data have been averaged to represent typical behaviour. However, this naturally disguises the pop-in events associated with film breakage seen on the individual curves.

4.2 Spherical Indenter

The load/partial-unload force-displacement results for the spherical indenter loading onto silicon and the aluminium film on silicon are shown in Figures 4(a) and 4(b) respectively. The data generated for silicon (Figure 4(a)) show very good reproducibility with less than 1nm variation in the penetration depths at maximum load. These results also show virtually one line indicating the behaviour to be entirely elastic in the load range (to 10 mN) investigated.

In contrast to the results on silicon the behaviour for indentations on the aluminium film show greater scatter (~10 nm) for the same maximum load. The results for an individual test in Figure 4(c) also show that whilst there is an initial elastic response for the first 0.5 mN the unload data show a departure from the load curve at heavier loads. There are also suggestions of steps on the individual curves suggesting sequential film cracking events in the oxide film as the spherical tip penetrates the aluminium film. Averaged data are shown in Figure 4d to represent typical behaviour and again the detail of the cracking events are lost.

5 DISCUSSION

Analysis of the results was performed on the basis that the behaviour is representative of a bulk homogeneous material rather than the oxide film lying on top of the aluminium film on the silicon substrate. Consideration of the individual responses of each layer will be given secondly. In the averaged plots shown in Figures 3(d) and 4(d) the machine compliance of 0.45 nm/mN has been removed from the data. Over the load range investigated this correction is not very significant.

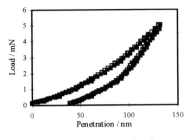

(a) Silicon – multiple plots

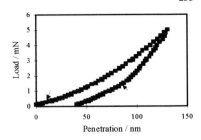

(b) Silicon – single test result, arrows indicate phase transformation

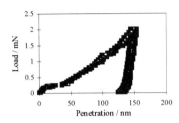

(c) Aluminium film on silicon – multiple plots

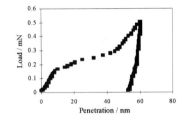

(d) Aluminium film on silicon – averaged data

Figure 3 *Load-penetration curves using the Berkovitch intenter*

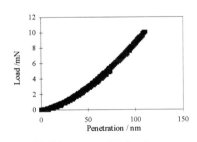

(a) Silicon – multiple plots

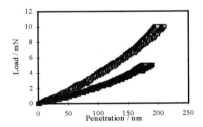

(b) Aluminium film on silicon – multiple plots

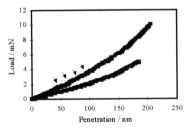

(c) Aluminium film on silicon – single test result, arrows indicate oxide film cracking

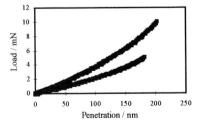

(d) Aluminium film on silicon – averaged data

Figure 4 *Load -penetration curves using the spherical indenter*

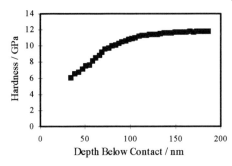

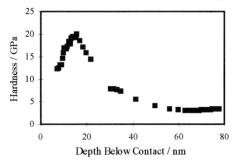

Figure 5a *Hardness vs depth for the Berkovitch indenter on silicon*

Figure 5b *Hardness vs depth for the Berkovitch indenter on aluminium*

The interpretation or analysis of the pointed indenter data from Figure 3 is plotted in Figure 5. In the case of the silicon (Figure 5a) the results show that the contact pressure or hardness rises to 12 GPa and remains constant throughout the load range investigated. This has been shown to correspond to the silicon phase transformation which occurs when a contact pressure of 11.25 GPa is exceeded in the hydrostatic compressive zone beneath the indenter[15,16]. The inflection along the unloading curve is typically observed for silicon (Pharr[17]) and from the analysis of Weppelmann et al[16] this is representative of the reverse phase transformation of silicon during unloading.

The results for the hardness of the aluminium film on silicon using the Berkovich indenter are shown in Figure 5b. In this case the hardness rises to an initial maximum before dropping to a much lower value and remaining nearly constant throughout the test. The lower values are between 3 and 4 GPa which is somewhat high for a pure aluminium metal indicative of some degree of impurity and partially reflecting the influence of the harder silicon substrate. Previous studies[18] of the behaviour of soft platinum films on silicon showed that considerable 'pile-up' formed about the indenter and that the hardness as a consequence was significantly overestimated because the location of the contact area was actually above the initial contact (i.e. material surface). It appears that the constraint imposed by the film being well bonded to the substrate causes a greater localised pile-up than for indentations only into a homogeneous substrate of aluminium. In the present case where a stiff and hard oxide layer covers the surface it is possible that the oxide layer causes an apparent increase in the size or shape of the indenter when loading the softer and more ductile aluminium metal underneath.

The interpretation of indentations using the spherically tipped indenter are shown in Figure 6. In

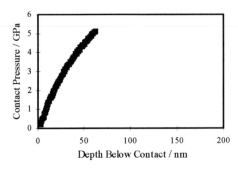

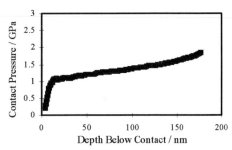

Figure 6a *Hardness vs depth for the sperical indenter on silicon*

Figure 6b *Hardness vs depth for the spherical indenter on aluminium*

the case of the silicon (Figure 6a) the behaviour is entirely elastic and the contact pressure monotonically rises to a maximum value of only 5.1 GPa. These results are consistant with those of Wepplemann et al[16] who showed that loads of ~25 mN were required to initiate the silicon phase transformation with a 5mm radius indenter.

The results for hardness or contact pressure for the aluminium film on silicon are given in Figure 6b and show that the contact pressure rises to a value of only 1.2 GPa before plateauing or rising slowly thereafter. Up to the value of 1 GPa the behaviour is entirely elastic and thereafter almost perfectly plastic except at deeper penetration where the influence of the substrate is being experienced. Again the influence of pile-up was ignored and this may have had a slight effect on the estimated hardness and modulus.

What are the causes of the difference in the results determined by the spherical and pointed indenters on the same substrate? The difference can be partially resolved by the location of the maximum shear stress beneath the contact and whether it is favourably located to initiate plastic flow of the more ductile phase. In the case of the ideal pointed indenter there is a stress singularity at the indenter tip which will either initiate plastic flow, brittle fracture or both. However, in reality all pointed indenters have a finite tip radius which may be in the range from 50–500 nm. With the presence of this tip radius the difference in behaviour of a nominally pointed and a deliberately rounded (sherically tipped) indenter is simply the scale and the load range in which they behave in a similar manner upon initial loading. The typical range for such behaviour is ~R/3, depending upon the cone or pyramid angle. The situation is slightly more complex in reality for a Berkovich indenter where only the first 20–30 nm of penetration behaves similarly to a spherical indenter. On the same basis an approximate estimate of the radius of the pointed indenter may be obtained from the initial penetration behaviour in silicon. These observations show that at ~0.2mN and penetrations of ~15nm the onset of the silicon phase transformation occurs (11.25 GPa[16]).

Using Equation {6} gives an estimate of the Berkovich tip radius as 0.375mm. An estimate of the contact circle radius is now possible for the initial penetration of the Berkovich indenter onto the oxide film on the aluminium. The contact pressure at the onset of the severe discontinuity may also be obtained in a similar manner using R = 0.375 mm and d » 8nm and F = 0.18 mN, which leads to a contact pressure of 19GPa. This value is comparable with the analysis of the contact pressure versus depth behaviour shown in Figure 5b. It is most likely at this threshold contact pressure that the brittle oxide film breaks and the softer aluminium metal beneath readily deforms with little resistance to the severe overload. Upon rapid penetration the simple analysis indicated a contact pressure of ~3 GPa which is somewhat high for aluminium metal and may be caused by the complex behaviour of this indenter now loading through the broken oxide 'islands' which would have a different geometry from the real indenter tip shape.

For the spherical indenter the maximum shear stress occurs below the oxide film and within the softer metallic layer. As such the 'plastic' penetration is indeed dominated by the metallic material rather than the brittle oxide film. For a Hertzian contact on an isotropic material the maximum shear stress occurs at a depth of 0.5 a below the centre of the indenter. For the more complex situation of a thin, high modulus layer on top of the softer metal coating the stress and strain fields are modified and the position of the maximum shear stress moves towards the surface. From FEM analysis of Djarbella et al[6] for a similar layer thickness and oxide/metal modulus ratio this occurs at 0.46 a. At the onset of 'plastic' deformation for the spherical indenter the contact radius may be estimated from the above and has a value of 0.26 mm which indicates that the maximum shear stress occurs at a depth of 130 nm, well below the oxide layer and within the aluminium metal. A more definitive analysis would require a complete FEM elastic-plastic modellisation of this deformation. The rôle of pile-up about the indenter is not so significant initially for a spherically tipped indenter because of the very small effective contact strains, or equivalent indenter angle (tan $\beta \approx a/R$, where β is the complimentary indenter angle).

6 CONCLUSIONS

The present study has compared the behaviour of a nominally pointed (Berkovich) indenter and a spherically tipped indenter during indentation of a 200nm thick aluminium film on silicon. The observations were able to deduce that there was an influential oxide film on the aluminium that significantly modified the initial loading response before it became dominated by the softer metallic material beneath. This was not as important for the case of the larger spherical tipped indenter because the location of the maximum shear stress occurred beneath the oxide film and so plastic deformation was not so significantly influenced by the brittle overlayer. Using the observations for both indenters loading on the silicon substrate it was possible to deduce a more critical estimate of the equivalent Berkovich indenter radius which was used to confirm the 'strength' and resistance to penetration of the oxide film and underlying metallic film upon initial penetration. A more rigorous analysis of this problem requires a more critical appreciation of the thickness of the oxide film on the aluminium and clearer indication of film break up so that an FEM model could be constructed. Nonetheless the micro-indentation technique has shown itself to be sensitive to changes in the properties of near-surface layers and can be used to explore the behaviour of other multilayer systems, implanted surface layers or functional gradient materials.

Acknowledgements

The authors wish to thank Prof. R. Novak of the Nagoya Institute of Technology for provision of the aluminium coated silicon sample.

References

1. N. G. Chechenin, J. Bottiger and J.P.Krog, *Thin Solid Films*, 1995, **261**, 219.
2. D. Tabor, 'Hardness of Metals', Oxford University Press, Oxford, 1951.
3. A. E. Giannakopoulos, P-L. Larsson and R. Vestergaard, *Int. J. Solids and Struct.*, 1994, **31**, 2679.
4. P-L. Larsson, E. Soderlund, A.E. Giannakopoulos, D.J.Rowcliff and R. Vestergaard, submitted to *Int. J. Solids and Struct.*, 1996.
5. N. Schwanger and F. Richter, *J. Adhesion and Tech.*, in press 1996.
6. H. Djarbella and R.D. Arnell, *Thin Solid Films*, 1992, **213**, 205.
7. H. Gao, C-H. Chiu and J. Lee, *Int. J. Solids Struct.*, 1992, **29**, 2471
8. J. Mencik, D. Munz, E. R. Weppelmann, E. Quandt and M.V. Swain, submitted to *J. Mater. Res.*, 1996.
9. M.V. Swain and J. Mencik, *Thin Solid Films*, 1994, **253**, 204.
10. M. Wittling, A. Bendavid, P.J. Martin and M.V. Swain, *Thin Solid Films*, 1995, **270**, 283.
11. W. Oliver and G. Pharr, *J. Mater. Res.*, 1992, 7, 1564.
12. J. S. Field and M. V. Swain, *J. Mater. Res.*, 1993, **8**, 297.
13. T. J. Bell, A. Bendeli, J. S. Field, M. V. Swain and E. Thwaite, *Metrologia*, 1991/92, **28**, 463.

14. A. J. Bushby, T. J. Bell and M. V. Swain, in preparation.

15 A. J. Bushby and M. V. Swain, "In Plastic Deformation of Ceramics", R.C. Bradt, C. A. Brooks and J. L. Roubort (Eds), Plenum Pub. Co., New York, 1995, p. 161.

16. E. R. Weppelmann, J. S. Field and M. V. Swain, *J. Mater. Res.*, 1993, **8**, 830.

17. G. Pharr, In: *"Thin Films: Stresses and Mechanical Properties III"*, W.D. Nix, J.C. Bravmann, E. Arzt and L.B. Freund, Eds., 1992.

18. J. Mencik and M. V. Swain, *MRS Symp. Proc.*, 1995, **356**, 729

1.4.4

Depth-Profile Elemental Concentration, Microhardness and Corrosion Resistance of Nitrided Chromium Steels

J. Flis[1], J. Mankowski[1], T. Bell[2] and T. Zakroczymski[1]

[1]INSTITUTE OF PHYSICAL CHEMISTRY OF THE POLISH ACADEMY OF SCIENCES, 01-224 WARSAW, POLAND

[2]DEPARTMENT OF METALLURGY AND MATERIALS, UNIVERSITY OF BIRMINGHAM, BIRMINGHAM, UK

1 INTRODUCTION

Benefits from the introduction of nitrogen to austenitic stainless steels include an increased stability of the γ phase, increased strength, hardness and wear resistance, and in some cases improved corrosion resistance. This caused a considerable interest in nitrogen alloying of these steels[1-5]

Nitrogen in solid solution increases resistance to pitting corrosion[6-9]; however, precipitation of nitrogen in the form of chromium nitrides causes a deterioration of the corrosion resistance due to chromium depletion of the matrix.[10-11] An adverse effect of nitriding on corrosion resistance has been observed particularly for diffusion zones of nitrided chromium steels[11-14]. Better corrosion resistance was observed for nitride phases of γ'[11], ε or ε + γ'[{15}], and especially for the so called "S phase"[11,16]. The S phase has a fcc structure similar to the nitrogen expanded austenite phase[11,16-20], being a nitrogen supersaturated solid solution with approximately 33 at.% N[20]. The corrosion behaviour and hardness of nitrided chromium steels have been related to the chemical composition and structure of the nitrided layers[11-15], nevertheless, some observations have not yet been fully explained. For example, it is not clear, why the transition region between the nitrided layer and matrix is more resistant to Marble's reagent than the layer above it[21], It is possible that the increase in corrosion resistance in this region might be associated with subtle changes in chemical composition and/or structure which cannot be detected by X-ray diffraction.

In the present work the depth profiles of the elemental concentration in nitrided layers were determined by glow discharge spectrometry (GDS) and Auger electron spectroscopy (AES) in an attempt to correlate the microhardness and corrosion behaviour with the chemical composition throughout the layers.

2 EXPERIMENTAL PROCEDURE

2.1 Experiments were conducted on three steels:

i. Austenitic Cr18Ni9Ti steel (type AISI 321) (C 0.05, Si 0.22, Mn 1.44, P 0.039, S 0.018, Cr 18.20, Mo 0.19, Ni 8.62 and Ti 0.26 wt.%); samples in form of 3 mm thick discs, 20 mm in diameter, were solution annealed at 1050°C and polished with wet emery papers to a 600 grit finish. Plasma nitriding was performed in a gas mixture of 25 vol.%N_2 + 75 vol.-%H_2 or 80 vol.-%N_2 + 20 vol.%H_2 under a pressure of 670 Pa at 585±10°C or 440°C for 6 h or 16 h.

ii. Martensitic Cr17Ni2 steel (type AISI 431) (C 0.17, Si 0.40, Mn 0.57, P 0.020, S 0.023, Cr 16.35 and Ni 1.95 wt.%); before nitriding, samples were oil quenched from 1000°C and then tempered for 2 h at 620°C, which resulted in a ferritic structure. Plasma nitriding was similar as given above.

iii. Sorbitic Cr3Mo steel (C 0.41, Cr 3.00, Ni 0.09, Mn 0.68, Si 0.33, P 0.026 and Mo 0.64 wt.%); samples were oil quenched from 920°C and then tempered for 2 h at 620°C. Plasma nitriding was performed at 540°C.

2.2 Elemental Depth Profile Analysis

Concentration profiles were obtained by glow discharge optical emmision spectroscopy (GDS) , using GDS-750 QDP equipment supplied by Leco Ltd. Depth profiles were also determined by AES on the surface of craters produced by grinding with a 41.4-mm diameter ball and diamond paste.

AES analysis was carried out on ground surfaces before and after a corrosion test. The spectrometer was operated at a beam voltage of 3.0 kV. Sample surface was cleaned by Ar^+ sputtering at 2.0 kV.

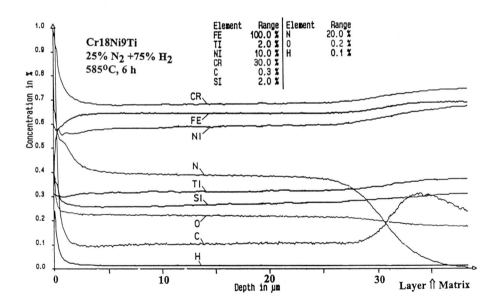

Figure 1 *GDS concentration depth profiles across a plasma nitrided layer in Cr18Ni9Ti steel treated in 25% N_2 + 75% H_2 at 585°C for 6 h. Arrow ⇑ denotes the depth of the nitrided layer/matrix interface*

2.3 Microhardness Corrosion Test

Microhardness was measured on cross sections using a Vickers tester with a load of 100g. The corrosion behaviour was examined at 25°C in non-deaerated 0.05 M Na_2SO_4 solution, acidified with H_2SO_4 to pH 3.0. Polarisation curves were measured at a potential scan rate of 1 mV s^{-1}, starting from a potential ~ 50 mV more negative than the open circuit value. Potentials were measured against the sulphate electrode, but in this paper they are referred to the saturated calomel electrode (SCE). Measurements were made on as-nitrided surfaces and at various depths obtained by grinding.

3 RESULTS AND DISCUSSION

3.1 Depth Profiles of Elemental Concentration

The concentration profiles obtained from GDS are shown in Figures 1–5 for steels after plasma nitriding under various conditions. For thick layers the GDS analysis was performed

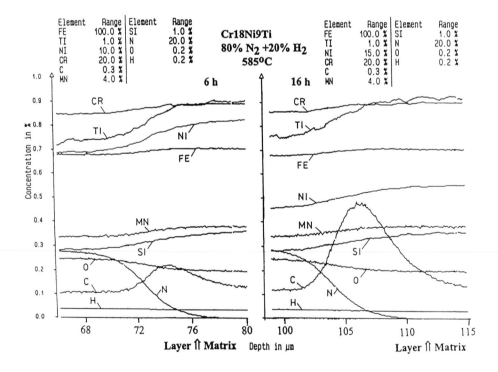

Figure 2 *GDS depth profiles across transition regions from nitrided layers to matrix in Cr18Ni9Ti steel treated in 80% N_2 + 20% H_2 at 585°C for 6 h and 16 h*

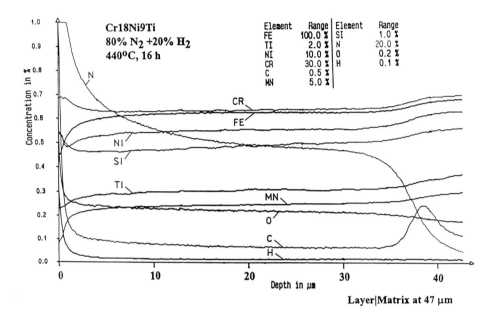

Figure 3 *GDS depth profiles across a nitrided layer in Cr18Ni9Ti steel treated in 80% N_2 + 20% H_2 at 440°C for 16 h*

in several stages with consecutive mechanical grinding. The depths indicated by the GDS equipment differed from the depths determined metallographically on cross sections through the craters, therefore, they were corrected. The depth of the craters was measured on the cross sections under an optical microscope, and it was compared with the depth given by GDS. The metallographically measured depth was assumed to be correct, and the GDS values were adjusted accordingly. The depths in Figures 1–5 present the "correct" values. Metallographically was also determined the distance from the surface to the interfaces of the nitrided layers into the matrix; in the Figures this distance is denoted by arrows ⇑.

Figures 1 and 3 present the profiles for the entire thickness of the nitrided layers, whereas Figures 2, 4 and 5 show the profiles only for the transition regions from the layer to the matrix (Figures 2 and 4) or from the compound to the diffusion zone (Figure 5).

The outermost surfaces (Figures 1 and 3) were enriched in Cr, N, O, C, and H. This enrichment in Cr and N can indicate the formation of a thin layer of chromium nitride on the surface. The enrichment in C, O, and H can occur both during the nitriding process (segregation of carbon from the bulk of metal, oxidation by trace amounts of oxygen, entry of hydrogen from the N_2/H_2 gas mixture) and during a subsequent contact with air (adsorption of CO_2, oxidation, and adsorption of water, respectively).

The elemental composition of the nitrided layer observed on the Cr18Ni9Ti steel after nitriding at 585°C (Figure 1) was almost uniform for the depths from about 3 mm down to the transition region. After nitriding at 440°C in the gas mixture with 80% N_2 (Figure 3), the concentration

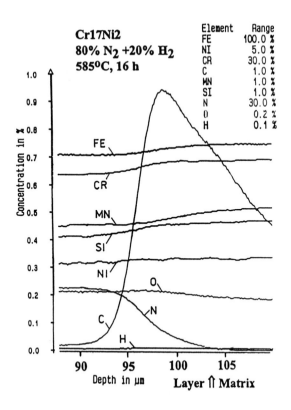

Figure 4 *GDS depth profiles across transition region from nitrided layer to matrix in Cr17Ni2 steel treated in 80% N₂ + 20% H₂ at 585°C for 16 h*

Table 1 *Depth of transition regions in plasma nitrided steels, concentration of carbon at the carbon peaks, concentration of nitrogen before the peaks, and C/N ratios*

Steel	Nitriding	Depth (μm)	C (wt.%)	N (wt.%)	C/N ratio
Cr18Ni9Ti	25% N₂ 585°C, 6h	35.0	0.090	7.55	0.012
Cr18Ni9Ti	80% N₂ 585°C, 6h	76.5	0.067	5.33	0.012
Cr18Ni9Ti	80% N₂ 585°C, 16h	112.0	0.140	5.55	0.025
Cr18Ni9Ti	80% N₂ 440°C, 16h	47.0	0.122	9.33	0.013
Cr17Ni2	80% N₂ 585°C, 16h	103.5	0.955	6.67	0.143
Cr3Mo	25% N₂ 540°C, 6h	6.8	0.278	10.22	0.027
Cr3Mo	80% N₂ 540°C, 6h	17.5	0.544	10.22	0.053

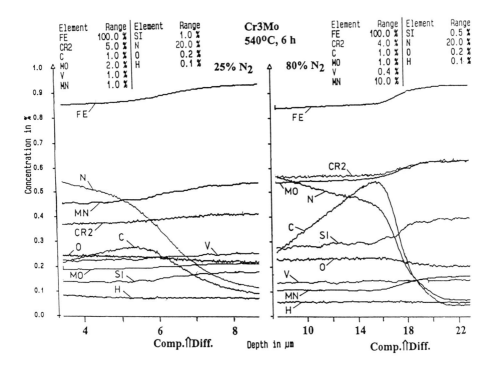

Figure 5 *GDS depth profiles across transition regions from compound to diffusion zones in Cr3Mo steel treated at 540°C for 6 h in N_2/H_2 gas mixtures with 25% N_2 and 80% N_2*

of nitrogen was significantly higher than that after nitriding at 585°C in the mixture with 25% N_2; at the outer surface it was close to 20 wt.% and it strongly decreased with the depth in the outer region. Probably, nitrogen was in a supersaturated solid solution in the austenitic lattice, as shown by other authors[11, 16-20]. The nitriding at 440°C produced a nitrided layer with a good corrosion resistance throughout the entire depth[22].

In the transition regions, which are denoted by the steep decrease in the nitrogen concentration, there always occurred a peak of carbon concentration. In the Cr18Ni9Ti and Cr17Ni2 steels the maximum of the carbon peak appeared in the region of the declining nitrogen concentration (Figures 1–4), whereas in the Cr3Mo steel the carbon maximum occurred at the beginning of the nitrogen decline (Figure 5).

The height of the carbon concentration peaks depended on many factors. Table 1 shows concentrations of carbon at the peaks, concentrations of nitrogen at a distance a few micrometers before the nitrogen decline, and the ratios of these concentrations (C/N ratios). In the Table there are also given the depths of the layer/matrix interfaces (for the Cr18Ni9Ti and Cr17Ni2 steels) or the compound/diffusion zone interfaces (for Cr3Mo).

Table 1 indicates that for the Cr18Ni9Ti steel the C/N ratio depended mainly on the time of

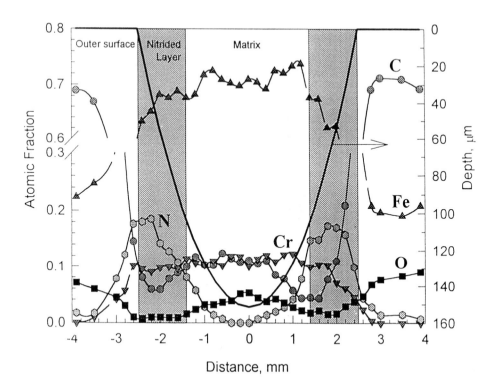

Figure 6 *AES depth profiles along diameter of a crater in Cr18Ni9Ti steel nitrided in 80%*
N_2 + 20% H_2 at 585°C for 16 h. Abscissa presents distance along the crater diameter;
parabola shows depth below the outer surface

nitriding: it increased by about two-fold with the extension of time from 6 h to 16 h. For the
Cr17Ni2 steel the C/N ratio was by about six times higher than for Cr18Ni9Ti; this might be
due both to the higher carbon content and to a faster diffusion in the ferritic Cr17Ni2 steel. In
the Cr3Mo steel, the C/N ratio rose with the increase in the N_2 content in the nitriding gas
mixture. Generally, the C/N ratio increased when transport of carbon was facilitated (longer
times, faster diffusion in ferritic structure), when carbon content in a steel was high, and when
nitrogen content in the gas mixture was high. Accordingly, the observed accumulation of
carbon can be explained by the displacement of carbon by nitrogen.

Accumulation of carbon in the near-substrate region of compound layers was earlier reported
by Haruman et al.[23] for plasma nitrocarburised iron-carbon alloys and low alloy steels; cementite
was found in this region.

3.2 AES Analysis

Figures 6 and 7 show the concentration profiles along the diameters of the craters developed
in the Cr18Ni9Ti and Cr17Ni2 steels, plasma nitrided in a 80% N_2 + 20% H_2 gas mixture at
585°C for 16 h. Atomic concentration C_x was assessed using the relation:

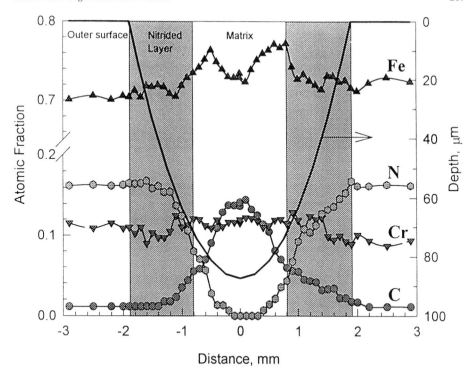

Figure 7 *AES depth profiles along diameter of a crater in Cr17Ni2 steel nitrided as given in Figure 6. The outer surface was ground to the depth of a few micrometers*

$$C_X = \frac{I_x}{S_x \sum \dfrac{I_i}{S_i}} \qquad \{1\}$$

where I_x is the peak height of element X, and S_x is a relative elemental sensitivity factor taken from[24]. The summation was made over one peak per element.

The thickness of the nitrided layers was determined metallographically, whereas distance from the outer surface (depth) at the crater walls was estimated from geometry of the craters. The depth is depicted in Figures 6 and 7 by parabola, whereas distance along the crater diameter is given on the abscissa. In this presentation the AES profiles do not exhibit any carbon peaks in the transition regions; however, the peaks may be hidden in the regions designated in Figures 6 and 7 as "Matrix". The occurrence of the carbon peaks may be manifested by the increased carbon concentration in the "Matrix" regions. This is probable especially for Cr17Ni2 steel (Figure 7), for which the carbon peak was particularly high (Figure 4 and Table 1).

In the used presentation of the AES profiles, the decrease in the nitrogen concentration in the nitrided layers with distance from the surface was more clearly seen than on the GDS

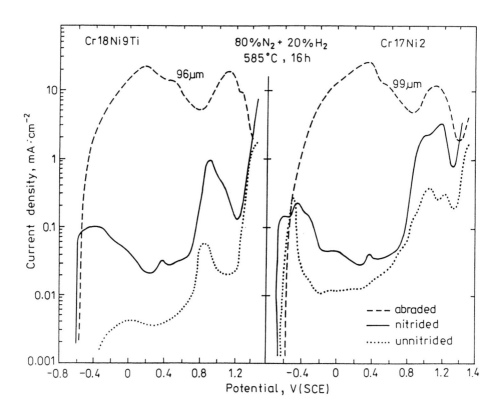

Figure 8 *Anodic polarisation curves for unnitrided and nitrided Cr18Ni9Ti and Cr17Ni2 steels in 0.05 M Na₂SO₄ + H₂SO₄ at pH 3.0 after abrasion to given depths*

profiles. Figure 6 shows that the concentration of oxygen in the "Matrix" region was higher than that in the nitrided layers. This demonstrates that an oxide film forms on the steel matrix more easily than on the nitrided layers.

3.3 Anodic Polarisation Curves

Polarisation curves were measured on untreated steels and for various depths of nitrided steels. Figure 8 shows polarisation curves for Cr18Ni9Ti and Cr17Ni2 steels prior to nitriding (designated as "unnitrided") and after nitriding; the measurements were made on as-nitrided surface ("nitrided") and after abrasion to given depths ("abraded").

It can be seen that anodic currents for unnitrided and nitrided Cr18Ni9Ti steel were lower than those for Cr17Ni2 steel, demonstrating a better corrosion resistance of the former steel both in the untreated and nitrided state. The outer surfaces of the nitrided steels exhibited a

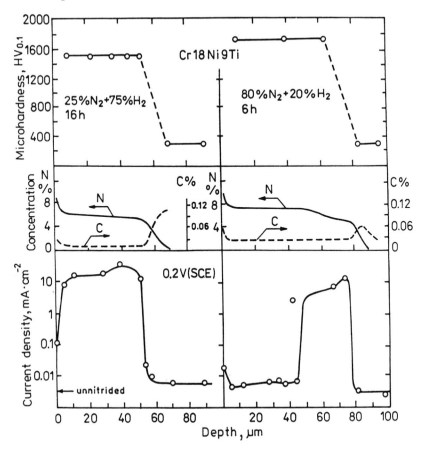

Figure 9 *Depth profiles of microhardness, of carbon and nitrogen concentration, and of anodic current at potential of 0.2 V (SCE) for Cr18Ni9Ti steel nitrided at 585°C in 25% N$_2$ +75% H$_2$ for 16 h, and in 80% N$_2$+20% H$_2$ for 6 h*

fairly good corrosion resistance; however, deep in the diffusion zones anodic currents were high and almost equal for the both steels. At these depths the diffusion zones did not passivate.

In the passive region the minimum current appeared at the potential of about 0.2 V (SCE). This current was used to represent the corrosion rate of nitrided layers as a function of distance from the surface.

3.4 Depth Profiles of Microhardness and Corrosion Rate

Figure 9 shows the depth profiles of microhardness, of the carbon and nitrogen concentration, and of anodic current at the potential of 0.2 V (SCE) for Cr18Ni9Ti steel nitrided at 585°C in 25% N$_2$ +75% H$_2$ for 16 h, and in 80% N$_2$+20% H$_2$ for 6 h. Microhardness dropped in the transition regions where the nitrogen concentration strongly decreased. The anodic current increased slightly at the end of the nitrided layers, and subsequently it decreased sharply by about three orders of magnitude in the region where nitrogen concentration started to decrease.

The anodic current profiles did not exhibit any peaks in the region of the carbon concentration maxima, so these profiles do not provide information on the effect of the carbon accumulation on the corrosion behaviour. It can be surmised that the accumulated carbon may increse the corrosion resistance rather than decrease it, because in the region of the carbon peak the current abruptly decreased, whereas before the peak the current slightly increased. The current drop was steeper than the nitrogen concentration drop, so possibly the abruptness in the current drop was due to the accumulated carbon and/or a nitrogen supersaturated solid solution; the latter is known to have an increased corrosion resistance[6-9]. A higher resistance of the transition region to chemical etching by Marble's reagent[21] may be also due to the carbon accumulation.

Plasma nitriding in 80% N_2+20% H_2 for 6 h produced a nitrided layer with two levels of nitrogen concentration. There occurred a distinct relationship between the anodic current and the nitrogen concentration levels: for the higher concentration in the shallower region of the layer the current was low and close to that for untreated steel, whereas in the deeper region the nitrogen concentration was lower and the current was by almost three orders of magnitude

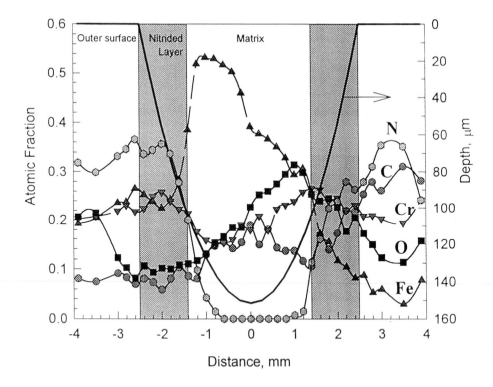

Figure 10 *AES depth profiles along diameter of a crater in Cr18Ni9Ti steel nitrided in 80% N_2 + 20% H_2 at 585°C for 16 h and held for 1.5 h in 0.05 M Na_2SO_4 +H_2SO_4 at pH 3.0 at the potential of 0.2 V (SCE)*

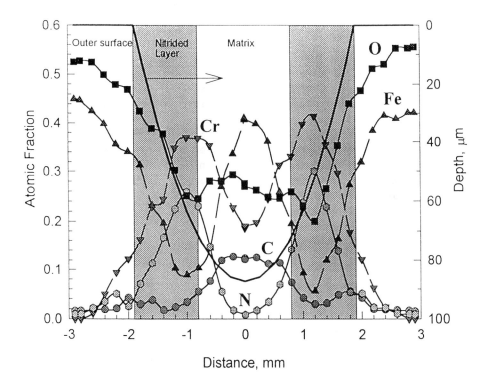

Figure 11 *AES depth profiles along diameter of a crater in Cr17Ni2 steel, treated as given in Figure 10*

higher. This indicates that the near-surface region can retain its ability to passivate even after nitriding at temperatures above the formation of the S-phase.

3.5 AES Analysis of Corrosion Products

Figures 10 and 11 show the AES elemental concentrations for the same steels as depicted in Figures 6 and 7, but after a   1.5h immersion in 0.05 M Na_2SO_4 +H_2SO_4 at pH 3.0 at the potential of 0.2 V (SCE) (minimum in the passive region, Figure 8).

Similarly as for the uncorroded surface of Cr18Ni9Ti (Figure 6), the concentration of oxygen on the matrix of this steel was higher than on the nitrided layer (Figure 10). On the matrix the rise in the oxygen concentration was parallel to the rise in the chromium concentration, indicating an enrichment in the chromium oxide or hydrated oxide.

For Cr17Ni2 steel the oxygen concentration was much higher than that on the Cr18Ni9Ti steel, demonstrating more intense deposition of corrosion products. In fact, the outer surface and the crater walls were covered with dark corrosion products; depth profiles for the oxygen and iron concentrations were similar, so these products might be composed mainly of iron oxides. Unlike in the case of the Cr18NiTi steel, for Cr17Ni2 steel the concentration of oxygen on the matrix surface was lower than that on the nitrided layer. This related to the intense corrosion of nitrided layers of the latter steel.

4 CONCLUSIONS

GDS depth profiles showed an accumulation of carbon in the transition regions from high to low nitrogen concentration in chromium steels plasma nitrided under various conditions. This accumulation was observed at the interfaces of the diffusion zone/matrix and of compound/ diffusion zone. It was facilitated by high average content of carbon and by its faster diffusion in the ferritic steels. The accumulation of carbon can result from the displacement by nitrogen.

Accumulation of carbon and/or the formation of a nitrogen supersaturated solid solution may be responsible for an increase in resistance to etching by Marble's reagent [21] and for a sharp decrease in anodic currents at the transition regions.

AES depth profiles along the crater walls were comparable with the GDS profiles. AES analysis showed that surface corrosion products on Cr18Ni9Ti steel were composed mainly of chromium oxides, whereas on Cr17Ni2 steel the products contained mainly iron oxides.

Acknowledgement

One of the authors (J. M.) gratefully acknowledges the COST Fellowship to perform the GDS analyses in the University of Birmingham.

References

1.   D.B. Rayaprolu and A. Hendry, *Materials Science and Technology*, 1988, **4**, 136.
2.   L.-A. Norstrom, *Mat. Sci.*, 1977, **11**, 208.
3.   J.E. Truman, *Stainless Steel Ind.*, 1978, **6**, 21.
4.   T. Bell and D. Kumar, *Metals Technol.*, 1978, **5**, 293.
5.   A. M. Ritter, M. F. Henry and W.F. Savage, *Metall. Trans. A*, 1984, **15**, 1339.
6.   M. Janik-Czachor, E. Lunarska and Z. Szklarska-Smialowska, *Corrosion* 1975, **31**, 394.
7.   J. E. Truman, M. J. Coleman and K. R. Pirt, *Brit. Corros. J.*, 1977, **12**, 236.
8.   Y. C. Lu, R. Bandy, C. R. Clayton and R. C. Newman, *J. Electrochem. Soc.* 1983, **130**, 1774.
9.   R. Bandy and D. Van Rooyen, *Corrosion* 1983, **39**, 227.
10.   A. V. Bilchenko and V. G. Permyakov, *Fiz. Khim. Mekh. Mater.* 1970, **6**, 6.
11.   Z. I. Zhang and T. Bell, *Surf. Eng.* 1985, **1**, 131.
12.   P. Sury, *Brit. Corros. J.*, 1978, **13**, 31.
13.   J. Flis, J. Mankowski and E. Rolinski, *Surf. Eng.*, 1989, **5**, 151.
14.   J. Mankowski and J. Flis, *Corros. Sci.*, 1993, **35**, 111.
15.   K. Ibendorf and W. Schröter, *Neue Hütte*, 1986, **31**, 333.
16.   P. A. Dearnley, A. Namvar, G. G. A. Hibberd and T. Bell, in "Plasma Surface Engineering" (Eds.: E. Broszeit, W. D. Munz, H. Oechsner, K.-T. Rie, and G. K. Wolf), DGM Informationsgesellschaft, Oberursel, 1989, Vol. **1**, p. 219.
17.   K. Ichii, K. Fujimura, and T. Takase, *Technol. Rep. Kasai Univ.*, 1986, **27**, 135.

18. S. P. Hannula, P. Nenonen and J. P. Hirvonen, *Thin Solid Films*, 1989, **181**, 343.

19. M. Samandi, B. A. Shedden, T. Bell, G. A. Collins, R. Hutchings and J. Tendys, *J. Vac. Sci. Technol.*, 1994, **B12**, 935.

20. K. Gemma and M. Kawakami, *High Temperature Materials and Processes*, 1989, **8**, 205.

21. K. Gemma, Y. Satoh, I. Ushioku, and M. Kawakami, *Surf. Eng.*, 1995, **11**, 240.

22. J. Mankowski and J. Flis, in 'Surface Engineering. Vol. I: Fundamentals of Coatings' (Eds.: P. K. Datta and J. S. Gray), Royal Society of Chemistry, Cambridge, Great Britain, 1993, p. 93.

23. E. Haruman, T. Bell, and Y. Sun, *Surf. Eng.*, 1992, **8**, 275.

24. E. Davis, N. C. MacDonald, P. W. Palmberg, G. E. Riach, and R. E. Weber, 'Handbook of Auger Electron Spectroscopy'. Physical Electronics Industries, Inc., Eden Prarie, MN, 1976.

1.4.5

Failure Modes in the Scratch Adhesion Testing of Thin Coatings

S. J. Bull

MATERIALS DIVISION, HERSCHEL BUILDING, THE UNIVERSITY, NEWCASTLE UPON TYNE, NE1 7RU, UK

1 INTRODUCTION

The scratch test has been used for some time to provide a measure of coating/substrate adhesion[1-6]. In the most commonly applied version of the test a diamond stylus is drawn across the coated surface under an increasing load until some well defined failure occurs at a load which is often termed the critical load, L_c. Many types of failure are observed which include coating detachment, through thickness cracking and plastic deformation or cracking in the coating or substrate[7-10]. In fact several different failure modes can occur at the same time and this can make results of the test difficult to interpret. Even so it has proved a useful method for assessing the adhesion of a range of hard coatings on softer substrates.

In general, for a harder coating on a soft substrate the spallation and buckling failure modes arise from interfacial detachment[8,9] and can thus be used as the basis for an adhesion test. Both modes are amenable to quantification and are discussed in some detail in this paper. The origin of these failure modes and the theoretical basis for analysing them is introduced, together with preliminary finite element results aimed at improving quantification.

2 SCRATCH TEST FAILURE MODES FOR HARD COATINGS

These can be broadly split into three categories:

1. Through thickness cracking – including tensile cracking behind the indenter[8,10], conformal cracking as the coating is bent into the scratch track[8,10] and Hertzian cracking[8]
2. Coating detachment – including compressive spallation ahead of the indenter[8,10], buckling spallation ahead of the indenter[8] or elastic recovery-induced spallation behind the indenter[8,11]
3. Chipping within the coating. The type of failure which is observed for a given coating/ substrate system depends on the test load, the coating thickness, the residual stress in the coating and the substrate hardness and interfacial adhesion. Generally the critical load at which a failure mode first occurs, or occurs regularly along the track, is used to assess the coating though there is a distribution of flaws and hence failures in most cases[6]. Comparisons between different coatings are only valid if the mechanism of failure is the same which requires careful post facto microscopical examination to confirm.

The adhesion related failures which are the basis of the scratch adhesion test are buckling and spallation and are described in more detail in the next section.

3 FAILURE MODES RELATED TO ADHESION

3.1 Buckling

This failure mode is most common for thin coatings (N < 10 μm) which are able to bend in response to applied stresses. Failure occurs in response to the compressive stresses generated ahead of the moving indenter (Figure 1). Localised regions containing interfacial defects allow the coating to buckle in response to the stresses and individual buckles will then spread laterally by the propagation of an interfacial crack. Spallation occurs when through-thickness cracks form in regions of high tensile stress within the coating. Once the buckle has occurred the scratch stylus passes over the failed region crushing the coating into the surface of the scratch crack formed in the substrate. Coating removal can be enhanced at this point or the failure may disappear completely depending on its size and the toughness of the coating.

Buckling failures typically appear as curved cracks extending to the edge of the scratch crack or beyond. They are often delineated by considerable coating fragmentation and have major crack planes perpendicular to the coating/substrate interface. In most cases buckles form in the region of plastic pile-up ahead of the moving indenter (Figure 1). The size of the buckle is typically less than or equal to the extent of pile-up. This would imply that the pile-up process controls the buckle failure mode to a great degree. This explains, to a large extent, the increase in critical load with substrate hardness for titanium nitride tool coatings on steel which is often reported[1] since in such coatings the buckle failure mode dominates. As the steel

(a)

(b)

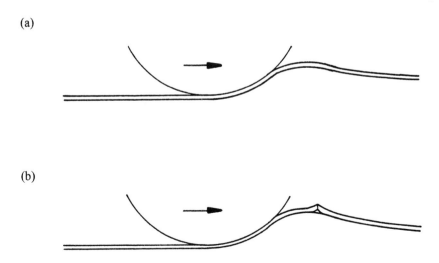

Figure 1 *Buckling failure mode in the scratch test; (a) pile-up ahead of the moving indenter and (b) interfacial failure leading to buckling. Through-thickness cracking results in removal of coating material*

hardness increases plastic pile-up ahead of the indenter is reduced and the bending stresses induced in the coating by the pile-up are limited. A higher normal load is needed to develop equivalent pile-up and bending stresses and thus the critical load increases. The correlation between buckle diameter and pile-up diameter is very close for oxide scales on MA956 or TiN on stainless steel (Figure 2). For TiN coatings on steel changes in buckle diameter can be produced by changes in interfacial structure and adhesion but within limits defined by the size of pile-up.

For thicker (>10μm) coatings where bending is less common the buckling failure mode is not observed. In fact the coating can suppress the formation of a narrowly defined pile-up region (Figure 3) and the stresses ahead of the indenter are less complex. Adhesive failure now occurs by a different mechanism. Initially compressive shear cracks form some distance ahead of the indenter through the thickness of the coating. These propagate to the surface and interface and generally have sloping sides which can act like a wedge. Continued forward motion of the indenter drives the coating up the wedge causing an interfacial crack to propagate. As the extent of interfacial failure increases the wedge lifts the coating further away from the

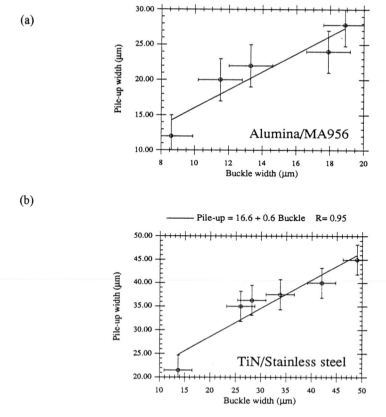

Figure 2 *Relationship between buckle size and the extent of pile-up in the scratch test (a) Alumina on MA956 (b) TiN on stainless steel 3.2 Wedge Spallation*

substrate creating bending stresses within it. Large enough displacements will cause a region ahead of the indenter to be detached in response to the tensile bending stresses created. When this happens the scratch diamond can drop into the hole left by removal of the coating (Figure 4) and there is a dramatic increase in scratch width and scratch depth. Pile-up is then often seen beside the track until the stylus climbs up the wedge and out of the hole. Whereas such failures are observed for alumina scales on MA956, much smaller failures are often produced for PVD TiN coatings.

4 EXPERIMENTAL

Samples of 304 stainless steel and the oxide dispersion strengthened alloy MA956 were cut into 20 x 10 x 2 mm sections, polished to a 1µm diamond finish, and cleaned and degreased in isopropyl alcohol prior to use. The stainless steel coupons were coated with TiN by sputter ion plating[12] at a temperature of 500°C and a bias voltage of –35V. Thicknesses in the range 1 to 15µm were deposited with a 120 nm titanium interlayer to promote adhesion. The MA956 samples were isothermally oxidised at temperatures between 1150°C and 1300°C for times up to 1400h to produce alumina scales up to 20 µm in thickness. The thickness of all coatings was measured by either ball cratering or metallographic cross sections.

(a)

(b)

(c)

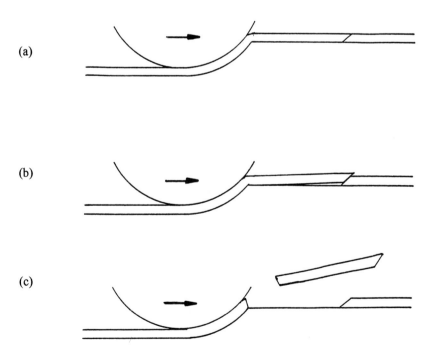

Figure 3 *Wedge spallation failure mode in the scratch test; (a) wedge crack forms some way ahead of the moving indenter; (b) continued forward motion drives the coating up the wedge opening up an interfacial crack; (c) through-thickness cracking close to the indenter leads to spallation*

Scratch testing was performed using a CSEM manual scratch tester fitted with a Rockwell 'C' diamond (200 μm tip radius). This is a dead-loaded machine where a separate scratch is made for each applied load. A 3mm scratch was made at each load. For the tests reported here scratches were made at 200 g intervals starting at 200 g. Care was taken to place the scratches sufficiently far apart that their deformation regions did not overlap. Critical loads for each failure mode were determined by post facto microscope examination of the scratch tracks. The critical load criterion used was the lowest load at which the failure occurred more than twice along the scratch track. Since the total number of wedge cracks produced was low it was not possible to perform a full Weibull statistics analysis[6].

Finite element modelling was used to assess the stress distribution for TiN coated stainless steel. A two dimensional plane strain model, which models the indentation as a cylinder rather than a sphere, was implemented in the DYNA3D code in order to achieve reasonable run times. The mesh was chosen to be symmetric about the *y* axis. The coating thickness was set at 2μm and reasonable materials properties were used for both substrate and coating. For the stainless steel substrate plastic deformation and work hardening was allowed, whereas for the TiN coating deformation is elastic up to fracture at ~500MPa. The indenter/coating friction coefficient was fixed at 0.15. Two models were run for comparison, a static indentation where the maximum vertical indenter displacement is 2 μm and a simulated scratch where the indenter is allowed to indent to 2 μm and is then moved tangentially 10 μm. Given uncertainties about the difference between cylindrical and spherical indentation, as well as questions about the quality of the materials data, the absolute stress values generated must be questionable. However, the difference between static indentation and scratches will be instructive.

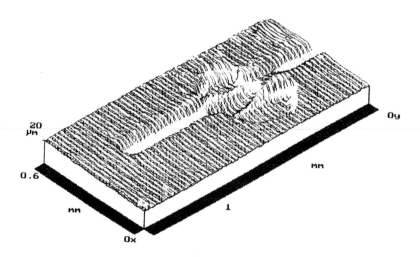

Figure 4 *3D stylus profilometry trace of a wedge spallation failure in 15μm alumina on MA956*

5 RESULTS

5.1 Scratch Test Failure Load Regimes

For all coatings investigated the critical load for buckle formation increases as the coating thickness increases (Figure 5). Wedge spallation does not occur until higher coating thicknesses and the critical load for wedge spallation decreases as thickness increases. According to Evans[13] the critical buckling stress symbol σ_b is given by:

$$\sigma_b = \frac{1.22E_c}{1-v_c^2}\left(\frac{t}{R}\right)^2 \qquad \{1\}$$

where E_c and v_c are the Young's Modulus and Poisson's ratio of the coating, t is coating thickness and R is the radius of the buckled region. This predicts that the critical buckle stress increases with coating thickness as is observed here. However, this equation assumes a planar interface which is not the case for the buckle failures observed in this study since the buckle is associated with pile-up. A complex stress state is expected which is confirmed by the finite element modelling. (see next section)

The wedge spallation failure mode depends on two distinct processes occurring[13]. Firstly a compressive shear crack must form and then interfacial detachment occurs. According to Evans the biaxial stress necessary to cause the wedge crack, σ_w, is given by:

$$\sigma_w = \sqrt{\frac{4E_c\gamma}{(1-v_c)\lambda}} \qquad \{2\}$$

where γ is the fracture energy and is the width of the wedge spalled region. The biaxial stress to produce the spall, σ_{sp}, after shear cracking has occurred is given by:

$$\sigma_{sp} = \sqrt{\frac{E_c\gamma_F}{(1-v_c)t}} \qquad \{3\}$$

where g_F is the interfacial fracture energy the total failure stress is given by the sum of these two. Since Equation 3 has a $1/\sqrt{t}$ dependence the critical load for wedge spallation is expected to decrease as coating thickness increases. This is what has been observed in this study. Clearly to characterise coating adhesion more fully it is necessary to determine the stresses responsible for coating detachment in terms of the critical load, L_c. The following sections describe how this can be achieved in some cases.

5.2 Finite Element Results

The main stress components in the coating at the coating/substrate interface have been extracted from the finite element data for both static indentation and scratching and are plotted in Figure 6. σ_{xx} (parallel to the surface) is tensile beneath the indenter in static indentation due to the bending of the coating as the substrate plastically deforms beneath it. At the edge of the

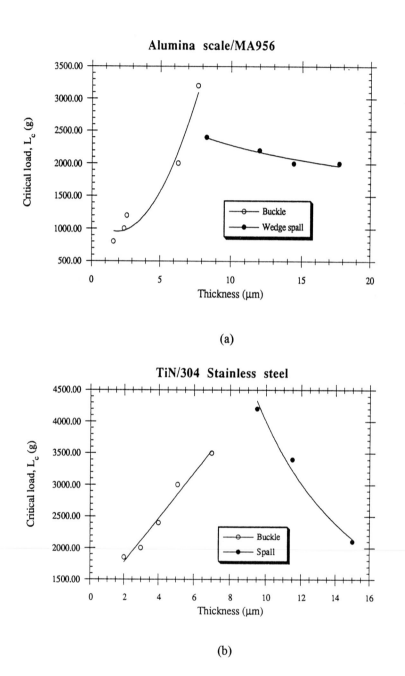

(a)

(b)

Figure 5 *Variation of critical load for wedge or buckle formation as a function of coating thickness for (a) Alumina on MA956 and (b) TiN on stainless steel*

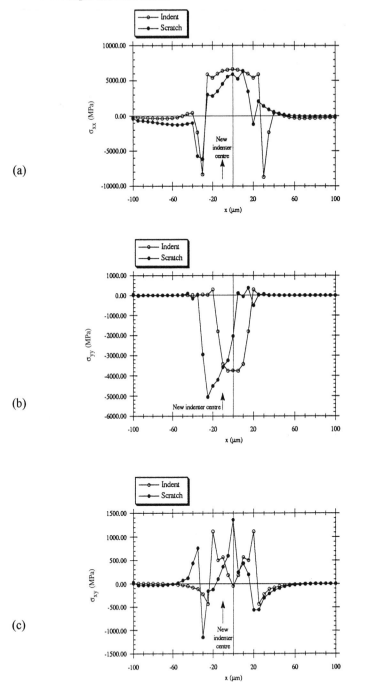

Figure 6 *Stress components in the coating next to the coating/substrate interface determined for static indentation and scratching using the Finite Element code DYNA3D (a) σ_{xx}, (b) σ_{yy} and (c) σ_{xy}*

contact the bending is in the opposite sense and compressive stresses are observed. This is exacerbated by pile-up. σ_{xx} quickly falls to zero outside the pile-up region. On moving the indenter the compressive stress is increased ahead of the indenter and reduced behind it, probably due to changes in the amount of bending in the coating. Well outside the contact region a compressive stress exists ahead of the indenter which approximates to a state of pure compression. In the region of bending at the edge of the contact the shear components σ_{xy} are also significant and those components at the leading edge of the indenter are increased by sliding. The stresses perpendicular to the interface σ_{yy} are compressive in the contact region as expected with tensile regions just outside in the bending zone. These tensile stresses are much reduced in the scratch case.

Clearly a very complicated stress state exists in the pile-up region close to the indenter where bending of the coating occurs. This is the region where buckling occurs and it makes the relationship between σ_b and L_c difficult to define. The stress state well ahead of the indenter where wedge cracking occurs is much simpler and a linear relationship between L_c and σ_F is expected for a given coating/substrate/indenter combination. This is amplified in the next section.

5.3 Quantification of Failure Stresses

The stresses responsible for coating detachment are a combination of the residual stresses remaining in the coating at room temperature, σ_R, and the stresses introduced by the scratch stylus, σ_s. Thus:

$$\sigma_F = \sigma_R + \sigma_s \qquad \{4\}$$

σ_R can be measured for both TiN and Alumina coatings by x-ray diffraction using the $\sin^2\psi$ method[14]. Table 1 contains the measurements made in this study.

The stresses induced by the indenter have been determined empirically. The critical load for coating detachment is known to decrease as the residual stress in the coating increases for a wide range of coatings such as TiN[15]. In the case of TiN coatings the residual stress can be increased by increasing the energy or flux of ion bombardment during deposition. Equating the change in scratch test critical load with the difference in measured residual stress enables

Table 1 *Residual stress at room temperature in the coatings measured by x-ray diffraction ($\sin^2\psi$ method).*

Coating/Substrate system	Residual stress (GPa)	
8μm Al$_2$O$_3$ on MA956		
Oxidised at 1150°C	3.75	±0.1
1200°C	4.04	±0.1
1250°C	4.11	±0.09
1300°C	4.24	±0.1
5μm TiN on stainless steel		
−30V bias	2.10	±0.20
−60V bias	6.03	±0.07

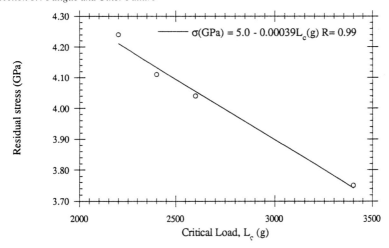

Figure 7 *Variation of critical load with residual stress for 8 μm alumina scales grown on MA956 at a range of oxidation temperatures*

a calibration factor to be determined. In this study 1g normal load in the scratch test equates to a 0.6 MPa compressive stress ahead of the indenter. For the alumina scales grown on MA956 the residual stress can be changed by altering the oxidation temperature. Plotting measured residual stress against critical load allows a calibration coefficient to be determined from the slope of the graph (Figure 7). For this material 1g normal load in the scratch test equates to 0.4 MPa. In both cases a linear relationship between critical load and stress is assumed which appears reasonable in these systems but further work is needed to determine the validity of the approach.

By plotting the calculated s_F against the reciprocal of the square root of coating thickness (Figure 8) it is possible to separate the two components contributing to wedging failure (Equations 2 and 3). The slope of this curve can be used to determine the interfacial fracture energy, γ_F, using the coating data in Table 1, whereas the intercept gives a measure of coating fracture strength, σ_w. These values are presented in Table 2.

In both cases the interfacial fracture energies are higher than that expected for the coating (~ 1 J/m^2) but lower than typical substrate values ($\sim 10^3$ J/m^2). This indicates that the failure crack is propagating at or near the interface with at least some crack tip plasticity occurring within the substrate. As γ_F increases the effective coating/substrate adhesion increases so the results here indicate that the TiN/stainless adhesion is better than that for alumina/MA956. Since the TiN coated stainless steel has much smaller spalled regions and the coating is considerably more abrasion resistant than the alumina scale the fracture energies are much as expected.

Table 2 *Compressive failure stress and interfacial fracture energy determined from the scratch test*

Substrate/coating	Interfacial fracture energy (J/m^2)	Wedge fracture stress (GPa)
304/TiN	451	2.52
MA956/Al$_2$O$_3$	16.9	4.1

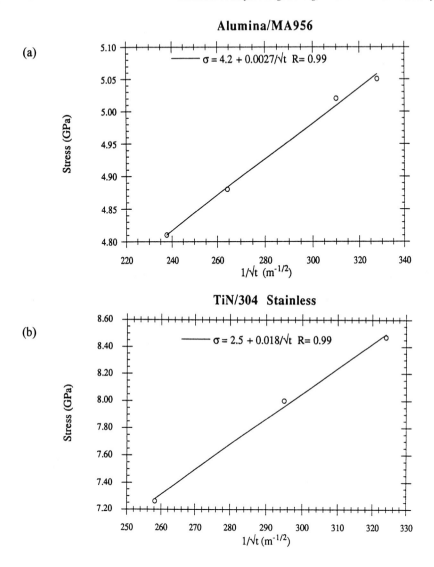

Figure 8 *Variation of critical failure stress, σ_F with the reciprocal of the square root of coating thickness for (a) Alumina on MA956 and (b) TiN on stainless steel*

6 CONCLUSIONS

The two main adhesion related failure modes in the scratch test are wedge spallation and buckling. Buckling occurs for thin coatings which are able to bend in response to applied stresses. The stresses responsible for failure are complex due to the fact that buckling is confined within the region of pile-up close to the indenter. For thicker, stiffer coatings wedge spallation becomes the dominant failure mechanism. This occurs well ahead of the moving indenter and the stresses which are responsible for failure approximate to a state of pure compression. Wedge spallation stresses can be quantified by calibration enabling an interfacial fracture energy to be determined. To derive the maximum benefit from the scratch test better theoretical models for the stress fields associated with a moving indenter in a coating/substrate system are needed.

Acknowledgements

This work was supported by AEA Technology and the CEC.

References

1. A. J. Perry, *Thin Solid Films*, 1983, **107**, 167.
2. P. A. Steinmann and H. E. Hinterman, *J. Vac. Sci. Technol.*, 1985, **43**, 2394.
3. J. Valli, *J. Vac., Sci. Technol.*, 1986, **A4**, 3001.
4. H. E. Hinterman, *Wear*, 1984, **100**, 381.
5. A. J. Perry, *Surf. Eng.*, 1983, **2**, 183.
6. S. J. Bull and D. S. Rickerby, *Surf. Coat. Technol.*, 1990, **42**, 149.
7. P. Benjamin and C. Weaver, *Proc. Roy. Soc. Lond., Ser A*, 1960, **254**, 177.
8. S. J. Bull, *Surf. Coat. Technol.*, 1991, **50**, 25.
9. S. J. Bull, *Materials at High Temp.*, in press, 1996.
10. P. J. Burnett and D. S. Rickerby, *Thin Solid Films*, 1987, **154**, 403.
11. R. D. Arnell, *Surf. Coat. Technol.*, 1990, **43/44**, 674.
12. D. S. Rickerby and R. B. Newbury, *Vacuum*, 1988, **38**, 161.
13. H. E. Evans, *Materials at High Temperature*, 1994, **12**, 219.
14. D. S. Rickerby, A. M. Jones and B. A. Bellamy, *Surf. Coat. Technol.*, 1989, **37**, 111.
15. D. S. Rickerby and S. J. Bull, *Surf. Coat. Technol.*, 1989, **39/40**, 315.

1.4.6
Strain Rate Sensitivity of Steel 03X26H6T

M. Fouad[1], T. M. Tabaghia[2] and S. Khalil[3]

[1]HEAD OF ROLLING DEPARTMENT, TIMS CAIRO

[2]CHAIRMAN, LIBYAN CORROSION SOCIETY, P. O. BOX 1281, BENGHAZI, LIBYA

[3]HEAD OF TIMS, CAIRO

1 INTRODUCTION

Large uniform plastic strains of superplastic materials (SPM) are derived from the high strain rate sensitivity of the flow stress[1]. In SPM where true strain hardening is minimal any neck which is present will always grow, although the rate of growth decreases with increasing strain rate sensitivity[1]. In general, the higher the strain rate sensitivity the greater the elongation to the failure although this rule is by no means universal[2].

The strain rate sensitivity (m) of SPM increases during the initial stages of superplastic flow if the deformation rate is initially less than the strain rate for maximum m-value. It is well known[2] that the actual magnitude of the increase in the strain rate sensitivity depends on interrelationship between strain rate, temperature and structure (both initial and final grain size and phase content). Then, the strain rate sensitivity index of SPM goes through a maximum with a definite value of strain rate (critical strain rate). It has been found that beyond a critical value of strain rate, the strain rate sensitivity decreases with increasing strain rate and hence superplasticity is slowly being lost due to competition from less rate sensitive process[2]. Generally for SPM, the high strain rate sensitivity is developed at commercially viable[3] strain rates in the range 10^{-4} to 10^{-2} l/s. The higher the strain rate in this range, the quicker will be the superplastic forming, minimising both forming times and deleterious effects (strain hardening due to grain growth) of microstructual evolution during deformation which have a marked effect on the strain rate sensitivity[4-6]. Thus, careful process control is required to confine forming to rather narrow range of temperatures and strain rates over which a given material exhibits a high strain rate sensitivity[7].

2 METHOD

Structure preparation route of stainless steel sheets 03X26H6T of 1.2mm thickness is given in study[8]. On the basis of tensile samples which were taken in longitudinal and transverse directions from these steel sheets the sigmoidal stress-strain rate relationships have been established[9] under temperatures of 900, 950 and 1000°C for holding times of 10^3 seconds (case 1) and 10^4 seconds (case 2) and sheet directions. The grain size (d) of steel 03X26H6T (steel A) in cases 1 and 2 is in the range of (1.588–2.459) and (2.309–4.090) m, respectively and the lower limits refer to a temperature of 900°C whereas the upper limits refer to a

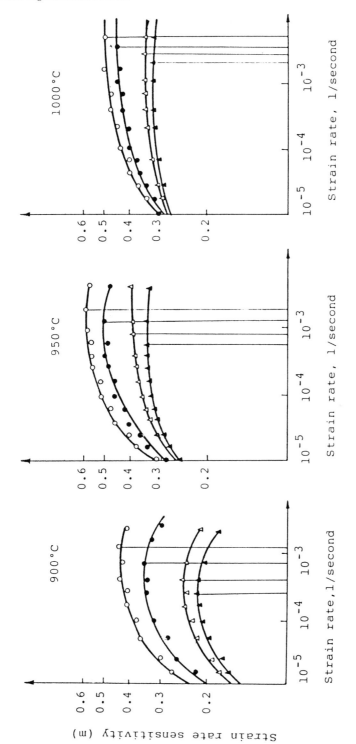

Figure 1 *The dependence of strain rate sensitivity on strain rate for steel 03X26H6T (average logarithmic relative error is 8.778%)*

temperature of $1000°C^9$. The stress-strain relationships (characteristics) of steel A were evaluated within the steady state and the slope of each characteristic, which is equal to the strain rate sensitivity, was noticed to increase with increasing strain rate up to a maximum value and beyond a critical value of strain rate the slope decreases with increasing strain rate, i.e. the strain-rate sensitivity index (m) of steel A goes through a maximum with strain rate. The m-values of steel A under different conditions are obtained from the slopes of the stress-strain rate characteristics of study[8]. Thus, the m-value of steel A was calculated for this study at different superplastic temperatures ($t = 900$–$1000°C$) and strain rates ($s = 10^{-5}$ to 10^{-2} l/s).

3 RESULTS ANALYSIS AND CONCLUSIONS

The values of strain rate sensitivity index (m) and their corresponding strain rates (s) are plotted on the logarithmic scale (log m – log s) for different cases (case 1 and case 2), sheet directions (rolling and transverse directions) and temperatures of 900, 950 and 1000°C. As it is well known that the strain rate sensitivity index (m) of steel A goes through a maximum with strain rate, the relationships between m an s are only given up until the maximum value of m is reached as given in Figure 1. The curves representing the plotted points with a relative error in the range of 3.92–4.99 are produced by means of the Smoothing Cubic Spline Funcrion (standard program) and are given in Figure 1. Thus, the curves of Figure 1 may be used for the prediction of (m) values at different conditions of strain rate, temperature and structure with high accuracy not less than 95%. Besides, under the abovementioned superplastic conditions it is found that the strain rate sensitivity is always higher in case 1 ($d = 1.588$–2.459mm) than that in case 2 ($d = 2.3089$–4.090mm) and in the rolling longitudinal direction than that in the transverse direction. These effects are also significantly dependent on the superplastic forming temperature and they increase as the temperature decreases from 1000 to 900°C. Thus, from analysis of Figure 1 it is easily concluded that steel A is anisotropic and its anisotropy increases as the strain rate increases in the range of 10^{-5} to 10^{-2} l/s, and the temperature decreases from 1000 to 900°C. It is also noticed from Figure 1 that maximum anisotropy of strain rate sensitivity of steel A takes place at the same conditions of achieving maximum m-values.

The peak curves (Figure 1) represent the maximum m-values of steel A at different conditions. The maximum m-values and their corresponding critical values of strain rate (s_c)

Table 1 *The maximum values of strain sensitivity indices (m) and their corresponding critcal strain rate*

Temperature t (°C)	Index	Strain rate s (l/s)		Anisotropy index of strain rate Ks ($K_s = s_l/s_t$)	Strain rate sensitivity index m		Anisotropy index of strain rate sensitivity K_m ($K_m = m_l/m_t$)
		Longitudinal direction s_l	Transverse direction s_t		Longitudinal direction m_l	Transverse direction m_t	
1000	Case 1	4.25×10^{-3}	3.16×10^{-3}	1.345	0.50	0.45	1.11
	Case 2	2.42×10^{-3}	1.91×10^{-3}	1.267	0.36	0.33	1.091
950	Case 1	1.84×10^{-3}	1.28×10^{-3}	1.438	0.58	0.50	1.160
	Case 2	8.32×10^{-3}	6.24×10^{-4}	1.333	0.38	0.34	1.104
900	Case 1	1.13×10^{-3}	6.67×10^{-4}	1.694	0.48	0.39	1.231
	Case 2	3.80×10^{-3}	2.51×10^{-4}	1.514	0.26	0.23	1.130

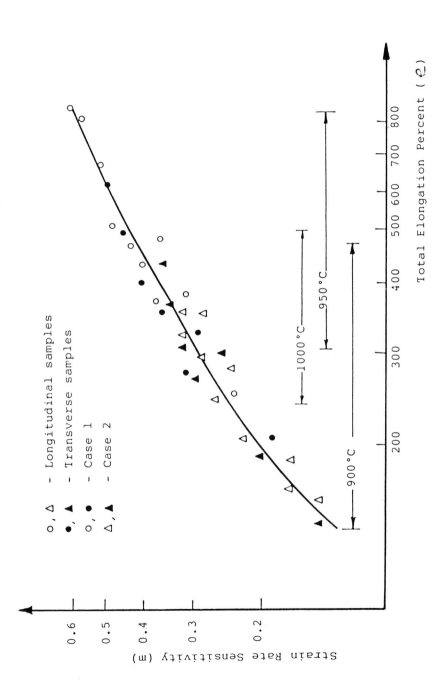

Figure 2 *The relationship between strain rate sensitivity (m) and the total elongation percent (e) of steel 03X26H6T*

at different superplastic conditions are given in Table 1. From the analysis of Table 1 it is found that the maximum values of strain rate sensitivity index (m) of steel A at different cases and in different sheet directions show a maximum at 950°C and minimum at 900°C. Besides, it is found that the peaks of curves (Figure 1) shift to the side of high strain rate values as the superplastic forming temperatures of steel A increases from 900 to 1000°C (Table 1). Moreover, from the analysis of Table 1 it is noticed that the maximum values of anisotropy indices (K_s and K_m) of strain rate and strain rate sensitivity respectively, are always higher than unity whereas, the values of K_s are always higher than that of K_m. It is also found from Table 1 that the values of K_s and K_m decrease as the temperature increases from 900 to 1000°C and this effect is more significant for K_s than that of K_m. The maximum values of anisotropy indices of steel A (K_s and K_m) take place at critical strain rates (s_c) which are corresponding to maximum values of strain rate sensitivity (Table 1). The transition from case 1 to case 2 leads to the increase of the critical values of strain rate (s_c) in the longitudinal and transverse directions by 30 and 26.67% ($t = 1000°C$), 34.48 and 32% ($t = 950°C$) and 43.18 and 38.46% ($t = 900°C$). Thus, it would be concluded for steel A that the change of its m-value is accompanied by a more remarkable change in its critical strain rate (s_c) and this effect is magnified as the superplastic temperature increases from 900 to 1000°C. This indicates that steel A is very sensitive to strain rate which is the main important characteristic of superplastic materials.

The elongation to failure (e) – strain rate (s) characteristics of steel 03X26H6T were obtained in study[10] and the critical values of strain rates (s_c) which correspond to the maximum values of elongation for this grade of steel were determined for the same superplastic conditions of this study. Thus, the values of elongation (e) and strain rate sensitivity index (m) of steel A which correspond to the same strain rates were plotted on the scale log m - log e (Figure 2) for different superplastic conditions of this study. The curve representing the plotted points with an average logarithmic relative error of 8.778% and correlation factor 0.987 is produced by means of the Smoothing Cubic Spline Function and is given in Figure 2. Thus within the previous accuracy the relationship log m - log e is independent of temperature, grain size and sheet direction. Besides, it is found from the analysis of Figure 2 that the patramaters m and e are strongly correlated and thus the strain rate sensitivity has a strong effect on the ductility of steel A. In general, the higher the m-value the grater the elongation to failure (Figure 2). It is observed from this study, and previous study[9] that the critical strain rates which correspond to the maximum elongation values occurred in a superplastic stainless steel 03X26H6T do not coincide with those which correspond to the maximum strain rate sensitivity values. The maximum elongation is encountered at a strain rate of an order of magnitude smaller than that for which strain rate sensitivity index is maximum. This could be attributed to the deformation induced grain growth which makes m strain dependent, distribution of cavity nucleating sites between tensile specimens which would significantly alter the total elongation and the variation in accuracy of machining specimens.

References

1. K. Osada et al., *Transactions of Iron and Steel Institute of Japan*, 1987, **27**, 713.
2. A. K. Ghosh and C. H. Hamiltin, *Metallurgical Transactions*, 1988, **13A**, 5, 733.
3. M. Foud, *J. Heat Treat. and Ther. Mech. Treat.*, 1991, **10**, 14.
4. M. Foud, *J. Heat Treat. and Ther. Mech. Treat.*, 1991, **9**, 20.
5. M. Foud, *J. of Ferrous Metallurgy*, 1991, **11**, 64.

6.   M. Foud, *J. of Metals of Academic Science Institute*, 1991, **6**, 77.

7.   M. Foud, *J. of Metal Forming Production*, 1991, **9**, 9.

8.   M. Foud, 'Stress Relaxation Testing of Stainless Steel 03X26H6T', 5th International Conference on Production Engineering, Design and Control, University of Alexandria, December 27–29, 1992.

9.   M. Foud, 'Anisotropy of Superplastic Stainless Steel 03X26H6T', 5th International Conference on Production Engineering, Design and Control, University of Alexandria, December 27–29, 1992.

1.4.7

Elasticity vs Capillarity, or How Better to Control Wetting on Soft Materials

M.E.R. Shanahan[1] and A. Carré[2]

[1]CENTRE NATIONAL DE LA RECHERCHE SCIENTIFIQUE, ECOLE NATIONALE SUPÉRIEURE DES MINES DE PARIS, CENTRE DES MATÉRIAUX P.M. FOURT, BP. 87, 91003 EVRY CÉDEX, FRANCE

[2]CENTRE DE RECHERCHE CORNING EUROPE, 7 BIS, AVENUE DE VALVINS, 77210 AVON, FRANCE

1 INTRODUCTION

The spreading of liquids on solid surfaces is of considerable interest and importance in many fields of activity. When inks, paints or varnishes are applied, they must wet their substrates before solidifying (by evaporation of the solvent or dispersion medium, curing or cooling). Similarly when applying a polymeric adhesive or in the manufacture of composite materials, adequate spreading is required to ensure good contact between two phases. Wetting also plays an important rôle in such techniques as spin casting, a process much used for producing thin films necessary in the electronics industry, and in many tribological applications. Apart from their uses in the industrial and commercial world, capillary phenomena are essential in many biological/medical phenomena such as blood circulation and eye irrigation, to name but two.

The basis of wetting phenomena is related to the fact that condensed matter, solid or liquid, usually presents a different structure at a surface or interface from that in the bulk, due to an asymmetry of the force field acting on local atoms or molecules. This leads to the concept of surface (or interfacial) free energy and the tendency, in the case of a liquid which is mobile, to minimise exposed surface. Surface tension is the force per unit length attempting to minimise the surface and, in the case of pure liquids and certain solids, may be taken as equivalent to surface free energy. (Strictly speaking, in the general case, the two quantities are different, but a discussion would be out of context here: we shall consider them as identical for present purposes.) Assigning the term γ_{ij} to an interfacial tension between two adjacent, immiscible phases i and j (solid S, liquid L and vapour V) and, by considering a force balance parallel to the (flat) solid surface, we obtain Young's equation for equilibrium[1]:

$$\gamma_{SV} = \gamma_{SL} + \gamma_{LV}\cos\theta_o \qquad \{1\}$$

where θ_o is the equilibrium contact angle subtended within the liquid phase between the solid surface and the tangent to the L/V interface at the line where the three phases meet (triple line). This simple derivation of Young's equation, although valid, is perhaps lacking a little in rigour. Nevertheless, the same result may be achieved using more involved analyses[2-4]. If we return to the simple force derivation, we may consider that if tensions equilibrate parallel to the surface, they must also do so in the normal direction. What balances $\gamma \sin \theta_o$ (where we define henceforth $\gamma = \gamma_{LV}$)? Most derivations assume implicitly that the solid surface under

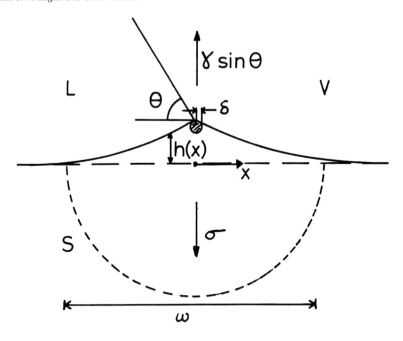

Figure 1 *Schematic representation of disturbed triple line region on a soft solid*

consideration is infinitely rigid, or in other words, that γ is insufficient to cause any disturbance. As will be seen below, this is often the case and the underlying assumption is adequate. However, in some cases, when the solid is sufficiently soft, a non-negligible deformation of the solid may result due to the component of γ normal to the (undeformed) surface and this may significantly modify wetting behaviour. This recent finding may have considerable impact in certain fields since spreading, under these circumstances, depends essentially on solid rather than liquid properties.

2 CAPILLARITY AND ELASTICITY AT THE TRIPLE LINE

We shall first of all consider the local behaviour near a solid/liquid/vapour triple line (N.B. the vapour could be a second liquid, immiscible with the first, but we shall refer to "vapour" in order to avoid possible confusion). Figure 1 corresponds to a schematic representation of the wetting front. Whilst horizontally (i.e. parallel to the undeformed solid surface), interfacial tensions balance, vertically, $\gamma \sin \theta$ is counteracted by a stress field within the solid leading to the formation of what we term as a "wetting ridge". Angle θ may either be the static equilibrium angle, θ_o, or a dynamic angle during spreading, $\theta(t)$ – the following argument also applies for the latter case since spreading phenomena will occur at a rate far below the speed of sound in the various phases and local force equilibrium is therefore assured.

More complete analyses have been effected[5,6], but we shall present here a simple scaling procedure in order to isolate the essential nature of the wetting ridge. Consider a zone of typical linear dimension ω in the vicinity of the triple line which corresponds to the solid disturbed by the capillary force $\gamma \sin \theta$. Within this region, a stress of order of magnitude a results. For local, vertical equilibrium to ensue, we have:

$$\omega \sigma \approx \gamma \sin \theta \qquad \{2\}$$

and adopting h as the height of the wetting ridge, we have local strain, ε, given in order of magnitude by:

$$\varepsilon = \frac{h}{\omega} \qquad \{3\}$$

Taking the solid to the Hookean, we have $\sigma = \varepsilon Y$ with Y being Young's modulus and using equations $\{2\}$ and $\{3\}$, we obtain as a scaling law:

$$h \approx \frac{\gamma \sin \theta}{Y} \qquad \{4\}$$

This simple expression really states that the perturbation to the solid is of dimension determined by the ratio of liquid surface tension to solid Young's modulus (i.e. capillarity vs elasticity) modulated by the angle of application of γ. Equation $\{4\}$ demonstrates why the vertical component of surface tension is often neglected. Typically, a "hard" solid such as a metal presents a value of Y of the order of 100 GPa and an organic liquid, a γ of the order of 3×10^{-2} Nm^{-1}. With a contact angle of 90°, $h \approx 10^{-2}$ Å : clearly a value without any physical meaning since we are well below the scale of applicability of continuum mechanics! However, many soft (and also dissipative – this is important in the following) solids such as gels and elastomers present Young's moduli of the order of 1 MPa (or less). With similar liquids, we may expect a value of h of about 30 nm or more – still small but non-negligible. As we shall see, this *mesoscopic* effect may lead to *macroscopic* consequences.

The above suggests a scale for h, but not the shape of the wetting ridge. More detailed analysis[6] leads to the form :

$$h(x) = \frac{2(1 - v^2)\gamma \sin \theta}{\pi Y} \ln \left| \frac{L}{x} \right| \qquad \{5\}$$

where v is Poisson's ratio for the solid, x is (horizontal) distance from the triple line and L is a cut-off distance which can be related to a datum depth within the solid (this constant is a usual consequence of two-dimensional stress analysis). Due to the logarithmic dependence of $h(x)$ on x, there is a cut-off at $|x| \approx \delta$. For $|x| < \delta$, equation $\{5\}$ no longer applies and the behaviour of the solid within this region is no longer linearly elastic[6]. Note from Figure 1 that the "true" contact angle is now somewhat greater than θ, but in the context of wetting, it may be shown that the conventional definition of this quantity, i.e. the angle subtended between the L/V and the *undisturbed* solid surface, remains the angle to be employed in subsequent analysis (unless we are dealing with an extremely small sessile drop (radius of the order of microns) and the solid is exceedingly soft[7]).

In the case of elastomeric solids, which we shall treat below, $v \approx 1/2$ and thus equation {5} simplifies to:

$$h(x) = \frac{\gamma \sin \theta}{2 \pi \mu} \ln \left| \frac{L}{x} \right| \text{ for } |x| > \delta \qquad \{6\}$$

where μ is the shear modulus of the elastomer $(Y = 3\mu)$.

3 DYNAMIC CONSIDERATIONS

Unless we are considering an exceedingly small sessile drop on a very soft solid, as stated above, the conventional contact angle, θ_o is still the pertinent value to be considered at equilibrium and the existence of a wetting ridge is rather academic. However, in the context of spreading (wetting or dewetting), the solid deformation can have a very marked effect on dynamic behaviour. As the triple line moves towards equilibrium, it is accompanied by a wetting ridge on a soft solid. The advancing wetting front raises the solid immediately ahead of its path and, after passing, releases the strained solid. As a result, a surface layer of solid (of thickness of the order of L as given by equation {6} typically of the order of 50 nm in the case of a soft elastomer) undergoes a strain cycle. If the solid were to be purely elastic, all strain energy would be restituted after passage of the triple line and the wetting ridge would not contribute to the overall dynamic behaviour. However, soft elastomers (and other soft solids) are, in general, viscoelastic and therefore lossy. As a result, the strain cycle results in energy loss by dissipation and this must be allowed for in the dynamic energy balance of spreading.

A rigorous approach to evaluating this dissipated energy has previously been undertaken[8] and later improved to allow for the behaviour of practical elastomers[9,10]. However, we shall present here a simplified argument in order to arrive at the essential scaling law and subsequently quote the more exact version.

Consider the work, E_1, effected when the vertical component of liquid surface tension, $\gamma \sin \theta$, "lifts" the local solid to a height, h, – that of the wetting ridge. This work, per unit length of triple line, is simply $\gamma h \sin \theta$ and using expression {4}, we obtain:

$$E_1 \approx \frac{\gamma^2 \sin^2 \theta}{Y} \qquad \{7\}$$

In fact, the horizontal component of liquid surface tension, $\gamma \cos \theta$, is not without effect, and leads to a stretching of the surface layer of solid on the vapour phase side of the triple line for $q < p/2$, or on the liquid side for $\theta > \pi/2$. By analogy with expression {7}, we have a second contribution to the work effected, E_2:

$$E_2 \approx \frac{\gamma^2 \cos^2 \theta}{Y} \qquad \{8\}$$

and thus the overall work supplied by the liquid surface tension and transferred to the solid (per unit length of triple line) is given by:

$$E \approx E_1 + E_2 \approx \frac{\gamma^2}{Y} \qquad \{9\}$$

As the wetting front advances at speed U, the local solid undergoes a strain cycle at a variety of frequencies, f, the actual frequency depending on the distance of the element of solid from the contact line at the moment under consideration. The solid the furthest from the contact line, yet still perturbed by the presence of the three-phase line, is at a distance of ca. ω and thus "feels" a strain cycle at frequency cm U/ω. At the other extreme, near the lower cut-off at $|x| = \delta$, the frequency is $\sim U/\delta$. The latter frequency will be dominant since it is in the direct vicinity of the three-phase line that the solid is strained the most. As a consequence, and using expression $\{9\}$, we may define the rate at which work is being done as:

$$\dot{E} = \frac{\gamma^2 U}{Y\delta} \qquad \{10\}$$

A viscoelastic solid may be characterized by a complex tensile modulus, $Y^*(f)$, a function of strain rate frequency, f

$$Y^*(f) = Y'(f) + iY''(f)$$

where Y' and Y'' are respectively storage and loss components of the modulus and $i = \sqrt{-1}$. The fraction of strain energy dissipated in a cycle is given by the loss tangent, tan $\Delta(f)$ (= $Y''(f)/Y'(f))$, or alternatively tan $\Delta(U)$ since $f \approx U/\delta$ in the present context. We may thus write an expression giving the quantity of energy dissipated per unit time and per unit length of triple line, Δ:

$$\dot{D} \approx \dot{E} \tan \Delta(f) \approx \frac{\gamma^2 U Y''(U)}{Y Y'(U)\delta} \qquad \{12\}$$

where $Y = |Y^*(f)|$.

Equation $\{12\}$ gives the basic scaling law for dissipation, but if a more rigorous treatment is applied, we obtain[8]:

$$\dot{E} \approx \frac{\gamma^2 U}{2\pi\mu\delta} \qquad \{13\}$$

Given that for a rubber $Y = 3\mu$, comparison of equations $\{10\}$ and $\{13\}$ shows that the simple scaling argument is in agreement with the more exact expression to within a factor of about 2!

However, use of the loss tangent in equation $\{12\}$ is an oversimplification. Instead, we shall make use of an empirical dissipation law obtained in the study of rate dependent adhesion of elastomers[11]. The following relation was proposed by Maugis and Barquins to explain the adhesion of elastomers in a variety of test configurations (peel, flat punch, etc.):

$$G - W_o = W_o \phi (a_T v) \qquad \{14\}$$

where G is strain energy release rate, W_o is the Dupré energy of adhesion[12], $\phi(a_T v)$ is a dimensionless function of separation rate v, and a_T is the Williams, Landel and Ferry time–temperature shift factor[13]. It was shown that $\phi(a_T v)$ varied as v^n where the constant n was of the order of 0.6. Given that similar elastomeric materials will be considered in wetting and dewetting, we replace equation {12} by equation {15}:

$$\dot{D} \approx \frac{\gamma^2 U}{2\pi\mu\delta} \left(\frac{U}{U_o} \right)^n \qquad \{15\}$$

where U_o is a characteristic triple line speed, depending on the solid (and probably to a lesser extent the liquid)[9,10].

4 TRIPLE LINE MOTION

The above description applies to energy dissipation during triple line motion, irrespective of the overall context of the process. In practical terms, the three-phase line can either advance as in the case of wetting, or recede as in dewetting. Both wetting and dewetting are of considerable importance. The former is essential in, for example, the manufacture of composite materials when fibres must be adequately (completely) covered by a polymeric resin. Dewetting is necessary in the application of windscreen wipers on a road vehicle! In the case of spin-casting or coating, the aim is to obtain an homogeneous thin layer of a liquid, before solidification, on a given substrate. If the equilibrium contact angle of the liquid, θ_o, is non-zero and the applied film is too thin, spontaneous dry patches may form and ruin the uniformity of the coating. This is an undesirable case of dewetting. In both cases of wetting and dewetting, it will be shown how deformation of a soft substrate may markedly slow down movement of the triple line.

4.1 Wetting – a Spreading Sessile Drop

4.1.1 Theory. We may consider the spreading of a sessile drop on a soft, lossy substrate rather like the advance of a "negative" crack and thus use fracture mechanics concepts as was the case in the derivation of equation {14} for the separation of an elastomer from a rigid solid. The term "negative" is used since the spreading of a drop leads to the creation of solid/liquid interface rather than separation.

Young's relation, equation {1}, refers to equilibrium at the triple line, but if the contact angle, as a function of time, $\theta(t)$, is greater than θ_o then there is a net spreading force acting per unit length of triple line and given by:

$$F = \gamma_{SV} - \gamma_{SL} - \gamma \cos \theta(t) = \gamma [\cos \theta_o - \cos \theta(t)] \qquad \{16\}$$

By analogy with equation $\{14\}$, this may be expressed as:

$$W_o - G = \gamma [\cos \theta_o - \cos \theta(t)] \qquad \{17\}$$

where:

$$W_o = \gamma_{SV} + \gamma - \gamma_{SL} = \gamma(1 + \cos \theta_o) \qquad \{18\}$$

and:

$$G = \gamma [1 + \cos \theta(t)] \qquad \{19\}$$

Thus the "strain energy release rate" is effectively an "instantaneous" value of Dupré's energy of adhesion with $\theta = \theta(t)$ instead of the equilibrium value ! The sign reversal in the left hand member of equation $\{17\}$, when compared to equation $\{14\}$ is simply due to the fact that we have a "closing crack" with a spreading liquid.

The equivalent of the right hand member of equation $\{14\}$, the dissipation, is composed of two additive terms. One is due to viscous shear within the liquid and the other due to viscoelastic losses in the wetting ridge. We may thus write:

$$W_o - G \approx \frac{3\eta\ell U}{\theta(t)} + \frac{\gamma^2}{2\pi\mu\delta} \left(\frac{U}{U_o}\right)^n \qquad \{20\}$$

The first term on the right hand side of equation $\{20\}$, due to viscous dissipation, involves liquid viscosity, η, and a logarithmic ratio of drop contact radius and a molecular cut-off, ℓ. This viscous term[14] is, however, small compared to the viscoelastic dissipation and we shall henceforth neglect it. Rearranging equation $\{20\}$, without the viscous contribution, gives :

$$U \approx U_o \left[\frac{2\pi\mu\delta}{\gamma^2}(W_o - G)\right]^{\frac{1}{n}}$$

$$= U_o \left[\frac{2\pi\mu\delta}{\gamma}(\cos \theta_o - \cos \theta(t))\right]^{\frac{1}{n}} \qquad \{21\}$$

4.1.2 Experimental. Several examples of viscoelastically controlled spreading have been observed[10], of which we shall present here the case of the liquid tricresyl phosphate (TCP) (Aldrich Chemicals Company Inc.), representing a non-volatile product of fairly low viscosity ($\eta = 7 \times 10^{-2}$ Pa. s) and average surface tension ($\gamma = 40.9 \times 10^{-3}$ Nm^{-1}), wetting a two component silicone rubber (RTV 630, General Electric Co., $\mu = 1.5$ MPa). Spreading on the elastomer will be compared to that of the same liquid on a rigid polymer (Teflon PFA, du Pont de Nemours, $\mu \approx 250$ MPa).

Small sessile drops $(2\mu$ l) were placed on the substrates and their contact angles, θ (t), followed optically as a function of time, t, until equilibrium, at 20°C. The solids were kept in a glass box in which the atmosphere was saturated in TCP vapour, both to prevent evaporation of liquid from the sessile drops and to reduce possible water pick-up from the atmosphere. Results of θ (t) as a function of t are given in Figure 2. It can be clearly seen that the kinetics of spreading to equilibrium is much slower on the soft, silicone elastomer than on the hard polymer. Spreading is complete on the hard solid after ~ 15 seconds whereas on the elastomer, the TCP drop is approaching equilibrium only after ~ 20 minutes. This unusual behaviour has been attributed to the existence of a wetting ridge. With the values of γ and μ quoted above, on the hard polymer, any wetting ridge should have a height of ~ 2 Å – this being an atomic dimension, we are really out of the range of continuum mechanics and such a ridge may reasonably be neglected. However, the height of the wetting ridge on the elastomer is ~ 30 nm – still small but non-negligible. In addition, the elastomer will be lossy whereas this is less likely for the rigid polymer.

Figure 3 shows an interpretation of the results of Figure 2, exploiting equation {21}, rearranged in the form :

$$\log [\cos \theta_{o} - \cos \theta (t)] \approx n\log U + \log\left[\frac{\gamma}{2\pi\mu\delta U_{o}^{n}} \right] \qquad \{22\}$$

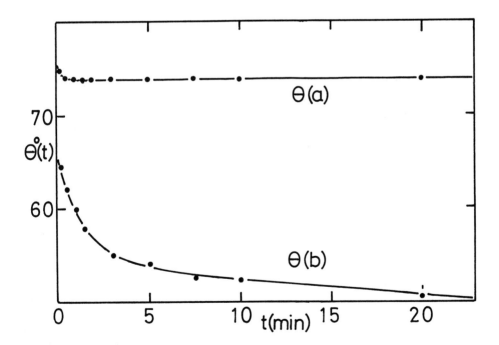

Figure 2 *Contact angle, θ (t), as a function of time, t, for 2 μl sessile drops of TCP spreading on (a) a rigid polymer, Teflon PFA and (b) a soft elastomer, RTV 630*

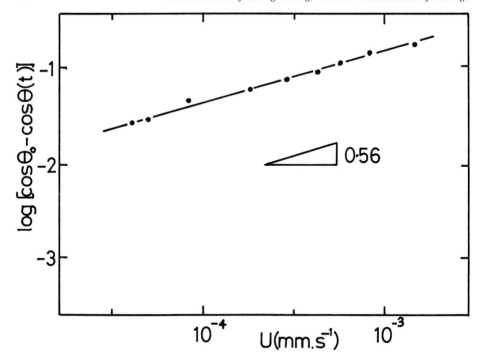

Figure 3 *Log [cosq$_o$ – cosq(t)] vs log U for a sessile drop of TCP on RTV 630. The gradient is equal to 0.56*

where the spreading speed, U, was calculated from the variation of drop contact radius with time, t. The linear relationship obtained is good evidence of the validity of equation {21} (or {22}). In addition, the value of the gradient, n, viz. 0.56, is in good agreement with the value of 0.6 obtained for viscous dissipation in adhesion experiments[11]. This similarity allows us to make an analogy between wetting and adhesion under dynamic conditions we have respectively a "wetting ridge" and an "adhesive ridge". Note that the gradient of the straight line of Figure 3 would be expected to be equal to unity in the absence of viscoelastic dissipation.

This general behaviour of viscoelastic braking has been observed with other solid/liquid pairs and, in fact, it has been shown directly that spreading may become viscosity independent[15]!

4.2 Dewetting – Hole Formation in a Liquid Film

The fact that spreading of liquids may be to some extent controlled by the bulk properties of the solid may be of some industrial use, but control of dewetting may be potentially more important.

4.2.1 Theory. Let us consider a circular "puddle" of liquid, L, on solid, S, in the presence of vapour, V, (Figure 4). The puddle is of radius R_0 and small initial thickness, e_0. We suppose that a small hole nucleates at the centre of the puddle and grows with radius $r(t)$ as time, t, passes because the equilibrium contact angle, θ_0, is non-zero: the liquid is unstable as a wetting film. The equilibrium thickness of a liquid film, e_c is given by[16]:

$$e_c = 2\sqrt{\frac{\gamma}{\rho_L g}} \sin\frac{\theta_o}{2} \tag{23}$$

where ρ_L and g are respectively liquid density (minus that of the surrounding fluid) and gravitational acceleration. If the film thickness, e_o, is less than e_c then it is unstable and holes, or dry patches, may nucleate spontaneously due to thermal agitation. We shall consider the regime in which $r(t) \ll R_o$ and for which $U = dr(t)/dt$ is relatively small (anticipating effects due to viscoelastic braking in hole formation). Assuming negligible liquid volatility, conservation of liquid volume leads to:

$$R_o^2 e_o = \left[R_o^2 - r^2(t)\right] e(t) \tag{24}$$

and thus:

$$e(t) \approx e_o\left[1 + \frac{r^2(t)}{R_o^2}\right] \approx e_o \tag{25}$$

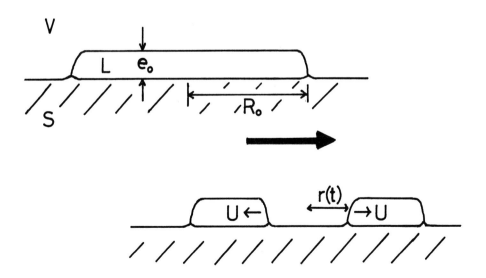

Figure 4 *Schematic representation of a circular wetting film on a soft substrate. A circular hole nucleates at the centre and grows with radius r(t) at speed U*

Here it is assumed that since U is small, displaced liquid will not collect in a rim around the hole: Laplace's pressure causes rapid equilibration (in contrast to rapid hole formation on rigid solids when the rim can be a determinant feature of the behaviour[16]).

From simple considerations of the local energy balance near the triple line, the driving force for dewetting per unit length of triple line, p, is given by:

$$p = \gamma + \gamma_{SL} - \gamma_{SV} = \gamma (1 - \cos \theta_o) \qquad \{26\}$$

if we assume the local contact angle to remain constant. In fact, dewetting is even closer than wetting to the analogous phenomenon of elastomer/rigid solid adhesion (opening "crack") and, comparing to equation {14}, we may write:

$$p = G - W_o \qquad \{27\}$$

where $G = 2\gamma$ and W_o is given, as before, by equation {18}. Thus, in the case of dewetting, the "strain energy release rate", G, is equivalent to the intrinsic work of cohesion of the liquid. Again neglecting viscous dissipation, we may equate $(G - W_o)$ to dissipation in the "dewetting" ridge and we obtain:

$$G - W_o \approx \frac{\gamma^2}{2\pi\mu\delta} \left(\frac{U}{U_o}\right)^n \qquad \{28\}$$

which may be compared directly with equation {20} for wetting, the sign on the left hand side being inverted since we are in a dewetting, rather than a wetting regime. Comparing the right hand members of equation {14} and equation {28}, we see that the former has an intrinsic term based on the energy of adhesion and a dissipative function implicitly containing the volume of elastomer involved in viscoelastic dissipation, whereas the latter has γ instead of W_o and the ratio γ/μ which is the approximate height of the (de)wetting ridge and thus a measure of the amount of elastomer involved in local energy dissipation.

Using equation {26}, we may rearrange equation {28} and integrate to obtain a more directly usable form, viz.:

$$r(t) \approx U_o \left[\frac{2\pi\mu\delta}{\gamma}(1 - \cos\theta_o)\right]^{\frac{1}{n}} t \qquad \{29\}$$

4.2.2 Experimental. In the example presented here, the same liquid and rigid solid have been used, viz. TCP and Teflon PFA, whereas for reasons of experimental convenience, a related, but different silicone elastomer was employed (RTV 615, General Electric Co. $- \mu = 0.8$ MPa).

Circular "puddles" of TCP of 50 mm diameter were obtained by attaching a 0.1 mm thick ring of plasticized adhesive paper on the solids and filling the inside with liquid. A thin layer was obtained by drawing across the liquid surface with a microscope slide. Calibration at 0.1 mm thickness was assured by the spacer. From equation {23}, it can be found that the equilibrium film thickness is of the order of 1.5 mm and thus films of 0.1 mm thickness are unstable, leading to spontaneous dry patch formation ($\theta_0 = 47°$ and $61°$, respectively for TCP on RTV 615 and Teflon PFA).

Dewetting experiments, at least in the early stages presently of interest, are much more rapid than wetting experiments and thus in the present work, observation was effected using a video camera mounted on a low power optical microscope.

Results are presented in Figure 5, in the steady-state regime, corresponding to hole radius, r(t), as a function of time, t[17]. Transient behaviour for t < 1 second is neglected. Results corresponding to dewetting on the hard polymer refer to direct measurement of triple line motion rather than observation of hole formation since, due to the high rate of dewetting, it was not found possible to observe the entire dry patch formation. In the case of dewetting on the silicone elastomer, it can be clearly seen that the hole radius is a linear function of time, in agreement with equation {29}, and the mean rate was found to be 0.39 ± 0.03 mms⁻¹. In addition, it was observed that no noticeable rim appeared near the receding triple line, thus corroborating our initial assumption that dewetting would be sufficiently slow for accumulation of liquid near the triple line not to occur. It was found that the average speed of dewetting on Teflon PFA was 8 ± 2 mms⁻¹. Thus, the dewetting rate is more than an order of magnitude lower on the lossy solid than on the (more or less) rigid substrate. Despite similar equilibrium wetting properties (θ_0 in the range 50–60°), the dynamic dewetting properties vary enormously as a result of the relative deformability of the substrates and the formation of a (de)wetting ridge which must travel with the triple line.

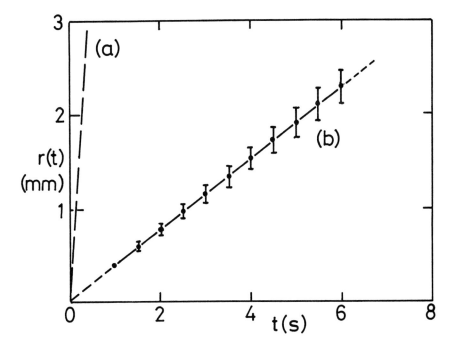

Figure 5 *Radius of spontaneously forming dry patch due to dewetting, r(t), as a function of time, t. (a) Expected behaviour on a hard polymer, Teflon PFA (obtained from direct observation of spreading speed rather than hole size) and (b) experimental values on a soft elastomer, RTV 615 (mean values of 5 experiments with standard errors)*

5 "ENGINEERING" WETTING, OR DEWETTING, RATES

When wetting rates are to be adjusted in an industrial context, such as paint application, or in lubrication, the usual approach is to modify the viscosity of the liquid in question by adding fillers, or thickeners. An everyday example of this need is the use of thixotropic properties in household paints. However, we can now see that there are potentially situations where solid substrate properties may be altered instead of those of the liquid. Indeed, there may well be cases where modification of the liquid properties are impossible – such would be the case in certain medical applications, for example catheters in contact with body fluids and contact lenses in contact with lachrymal fluid (tears !). Since this new "tool" for controlling wetting, or dewetting, rates is restricted to soft solids, we shall consider specifically the case of elastomers, although similar basic ideas may possibly be extended to such materials as gels.

We have seen in equations {12} and {15} that the basic viscoelastic dissipation term during triple line motion on an elastomer scales with the height of the wetting ridge $h \approx \gamma / Y$ (assuming for simplicity, a contact angle of $\pi/2$), or alternatively γ/μ. Let us consider the substrate shear modulus, μ. The classic theory of rubber elasticity tells us that:

$$\mu = \frac{\rho RT}{M_c} \qquad \{30\}$$

where r is the density of the elastomer, M, the average inter-cross-link molecular weight (in gm.mole^{-1}) and R and T are respectively the ideal gas constant and (absolute) temperature[18].

Considering the local gradient of the wetting ridge, from equation {6}, we have:

$$\left| \frac{d}{dx}.h(x) \right| = \frac{\lambda \sin \theta}{2\pi\mu|x|} \approx \frac{h}{|x|} \qquad \{31\}$$

and substituting for μ from equation {30},

$$h \approx \frac{\gamma M_c \sin \theta}{2\pi\rho RT} \equiv \frac{\gamma M_c \sin \theta}{2\pi\rho N_o k_B T} \qquad \{32\}$$

where N_o is Avogadro's number and k_B Boltzmann's constant. Now the volume of an inter-cross-link molecular chain is given by Nv^* where N is the number of elastomer monomer units and v^* the volume of a monomeric unit:

$$Nv^* = \frac{M_c}{N_o \rho} \qquad \{33\}$$

and thus we obtain:

$$h \approx \frac{N\gamma v^* \sin \theta}{2\pi k_B T} \qquad \{34\}$$

With $\gamma \approx 3 \times 10^{-2}$ Nm^{-1}, $v^* = 10^{-28}$m^3, $k_B = 1.38 \times 10^{-23}$ JK^{-1} molecule^{-1} and $T \approx 295$K, we find that $\gamma v^*/(k_B T)$ is of the order of 10^{-9} m, a length comparable to the diameter of a monomer in the elastic chain, d. We may thus conclude, in order of magnitude, that $h \sim Nd$. The height of a wetting ridge is thus comparable to the extended length of an inter-cross-link molecular chain in the elastomeric molecular structure. In principle, equation {31} is only valid for low values of the local gradient, or $|x| \gg h$. Nevertheless, the simple argument above gives an approximate relation between wetting ridge height and the degree of crosslinking of the elastomer.

Let us now consider the effect of cross-linking on spreading behaviour, in particular, in the case of dewetting when viscosity effects are negligible. From equation {31} and the above discussion, and allowing for the inclination of the L/V interface, we have:

$$h \approx \frac{\gamma \sin \theta_o}{2 \pi \mu} \approx Nd \sin \theta_o \qquad \{35\}$$

where we assume, for simplicity, that $\theta \approx \theta_o$.

Using expression {35}, as well as equations {26} and {27}, in equation {28}, we obtain, after rearrangement :

$$U \approx U_o \left[\frac{2 \delta \sin^2 \left(\frac{\theta_o}{2} \right)}{Nd} \right]^{\frac{1}{n}} \qquad \{36\}$$

Realising that typically $\delta \approx 5d$, and taking $n \approx 0.5$ (a value similar to that found experimentally), we find:

$$U \approx 100 U_o N^{-2} \sin^4 \left(\frac{\theta_o}{2} \right) \cong \frac{U_o}{M_c^2} \qquad \{37\}$$

Clearly the prefactor of 100 is only approximate and will depend anyway on the choice of liquid and elastomer. The main significance of relation {37} is to show the dependence of dewetting rate on (approximately) the inverse square of the intercrosslink molecular weight. Lower degrees of crosslinking will lead to slower dewetting. Note also from equations {36} and {37} that dewetting speed becomes independent of liquid surface tension, at least explicitly (it will be implicitly involved in the value of θ_o).

From the above, we may see that it should be possible to "engineer" the wetting or dewetting rates of liquids on elastomers. The above conclusion could also have been arrived at qualitatively simply by considering that the elastic modulus of an elastomer generally increases with degree of cross-linking, i.e. by decreasing M_c (equation {30}). In addition, higher degrees of cross-linking tend to decrease tan Δ (equation {12}) or increase U_o (equation {15}), thus lowering the lossy character of the material.

The example given is fairly simple, but materials engineers may be able to follow the general idea presented above and "tailor" soft solids in order to obtain a desired dynamic behaviour in liquid spreading.

6 CONCLUSIONS

Under classic conditions of wetting or dewetting, kinetic behaviour is governed by an energy balance invoking a capillary imbalance producing the motive power and viscous dissipation within the liquid restraining triple line motion. It has been shown that on a soft solid, the component of liquid surface tension acting perpendicularly to the (flat) solid surface leads to a local deformation, or wetting ridge. Although this wetting ridge is of minor importance under (most) static conditions, during spreading, or more precisely motion of the solid/liquid/vapour triple line, the fact that it must accompany the movement leads to energy dissipation. This dissipation can markedly affect the rate of triple line motion, to the extent that in some cases, it becomes independent of liquid viscosity ! Both wetting and dewetting phenomena can be slowed down by an order of magnitude or more. It is shown how the dimensions of the wetting ridge are related to the degree of crosslinking of the soft substrate in the case of an elastomer. Finally, it is suggested that by judicious choice of solid substrate properties, wetting or dewetting properties may be "engineered" or "tailored" to specific uses.

References

1. T. Young, *Phil. Trans. Roy. Soc.*, 1805, **95**, 65.
2. R. E. Collins and C. E. Cooke, *Trans. Faraday Soc.*, 1959, **55**,1602.
3. R. E. Johnson, *J. Phys. Chem.*, 1959, **63**, 1655.
4. M. E. R. Shanahan,'Adhesion 6', ed. K.W. Allen, Appl. Sci. Pub., London, 1982, p. 75.
5. G. R. Lester, *J. Colloid Sci.*, 1961, **16**, 315.
6. M. E. R. Shanahan and P.G. de Gennes, 'Adhesion 11', ed. K.W. Allen, Elsevier, London,1987,71.
7. M. E. R. Shanahan, *J. Phys. D: Appl. Phys.*, 1987, **20**, 945.
8. M. E. R. Shanahan, *J. Phys. D: Appl. Phys.*, 1988, **21**, 981.
9. A. Carré and M. E. R. Shanahan, *C.R. Acad. Sci.*, Paris, Série II, 1993, **317**, 1153.
10. M. E. R. Shanahan and A. Carré, *Langmuir*, 1995, **11**, 1396.
11. D. Maugis and M. Barquins, *J. Phys. D: Appl. Phys.*, 1978, **11**, 1989.
12. A. Dupré, Théorie Mécanique de la Chaleur', Gauthier-Villars, Paris, 1869, p. 369.
13. J. D. Ferry,'Viscoelastic Properties of Polymers', 2nd edition, Wiley, New York, 1970, p. 314.
14. P. G. de Gennes, *Rev. Mod. Phys.*, 1985, **57**, 827.
15. A. Carré and M. E. R. Shanahan, *Langmuir*, 1995, **11**, 24.
16. F. Brochard, C. Redon and F. Rondelez, *C. R. Acad. Sci.*, Paris, Série 11, 1988, **306**, 1143.
17. A. Carré and M. E. R. Shanahan, *Langmuir*, 1995, **11**, 3572.
18. L. R. G. Treloar, The Physics of Rubber Elasticity', Clarendon, Oxford, 1949, p. 187.

1.4.8

The Influence of the Electrodeposition Parameters on the Gas Sensitivity and Morphology of Thin Films of Tetrabutylammonium BIS(1,3-Dithiol-2-Thione-4,5-Dithiolate) Nickelate

J.R. Bates[1†], P. Kathirgamanathan[2], R.W. Miles[1].

[1]SCHOOL OF ENGINEERING, UNIVERSITY OF NORTHUMBRIA, ELLISON PLACE, NEWCASTLE UPON TYNE, UK

[2]DEPT. OF ELECTRONIC, ELECTRICAL AND INFORMATION ENGINEERING, SOUTH BANK UNIVERSITY, 103, BOROUGH ROAD, LONDON, UK

†Current address: HGa Ltd, Burderop Park, Swindon, SN4 0QD.

1 INTRODUCTION

There has been much demand recently for the development of cheap reusable gas sensor devices for both environmental and industrial applications. Toxic gas sensors and monitoring systems are widely available but no single system is suitable for all applications. Many of these devices utilise the fact that the electrical and optical properties of various metal oxides, such as tin oxide, change when exposed to certain gases[1]. The problem of achieving selectivity of response to a given gas or gases has still not been fully overcome, although the use of sensor arrays coupled with signal processing techniques has gone a long way towards realising this goal[2].

Metal oxide sensors are usually based on tin oxide although zinc and tungsten oxides have been investigated[1]. These sensors are based on the electrical conductivity of the metal oxide changing in different ambient atmospheres. Normally for a p-type material the conductivity is decreased by an oxidising gas and increased by a reducing gas (and conversely for n-type materials). These sensors are usually thin film or single crystal devices. The single crystal devices have not proved to be very successful because the conductivity changes involved are too small. Tin oxide thin film devices are available commercially for the detection of methane, hydrogen, carbon monoxide and hydrogen sulphide, although they need to be operated at temperatures of around 573K[1]. The non-specificity of thin film metal oxide sensors can be overcome to some extent by controlling the film microstructure and the use of incorporated catalysts. When these devices are operated at lower temperatures they often lose their sensitivity to many hydrocarbons and become sensitive to relative humidity.

The possibility of using organic conductors as gas sensitive materials has also been considered; the conductivity of metal phthalocyanines was shown to be sensitive to a number of different ambient gases in 1965[3]. Metal phthalocyanines are generally p-type semiconductors, so the capture of electrons by adsorbed gases from the semiconductor causes an increase in the conductivity of the film. The disadvantage with metal phthalocyanines is that to obtain a reasonable conductance, short response and recovery times, these devices need to be operated at temperatures in the region of 450K. Investigations into the effects of various gases on the

Table 1 *C, H, N, S analysis of various films*

Sample	C (%)	H (%)	N (%)	S (%)	Ni (%)
$Bu_4NNi(dmit)_2$	38.1	5.19	2.02	46.23	8.46
$(Bu_4N)_{0.29} Ni(dmit)_2$	24.39	1.99	0.77	61.58	11.27
$Ni(dmit)_2^0$	15.96	0	0	71.03	13.01
95:5	16.45	0.12	0.04	64.42	12.91
$Ni(dmit)_2^0{:}(Bu_4N)_{0.29}Ni(dmit)_2$					
90:10 Ni	16.94	0.23	0.09	69.94	12.80
$(dmit)_2^0(Bu_4N)0.29Ni(dmit)_2$					
Average film composition	16.91	0	0	69.60	-

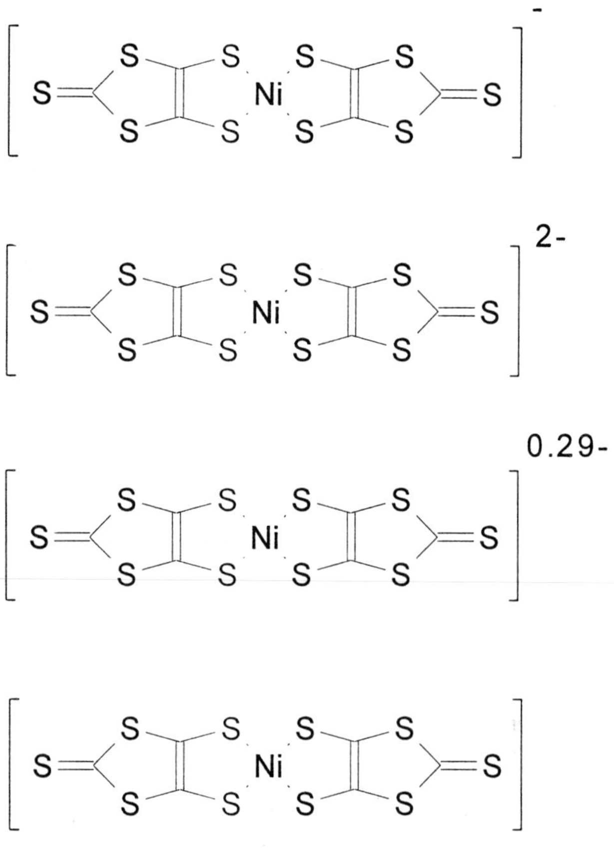

Figure 1 *Structure of $Bu_4NNi(dmit)_2$ showing the various valencies and room temperature conductivities*

electrical and optical properties of other organo-transition metal complexes have also been made[4,5].

Although these effects have been known for a considerable time, and have been used for the determination of charge carrier type in organic conductors, it is important to bear a number of things in mind when considering their use for gas detection: the magnitude of the conductivity change has to be such that it is easily measured; the rate of change needs to be sufficiently fast, and the change needs to be reversible. These three conditions are primarily governed by the strength of the charge tranfer interaction between the gas and the semiconductor. If this interaction is too weak, the effect on the activation energy for carrier activation is small, and if the interaction is too strong, it is localised and the charge carriers become coulombically bound to the adsorption sites. Furthermore, care is needed when interpreting the results: the device needs to calibrated in real conditions; although the device may respond at low analyte concentrations, it may be damaged at higher concentrations, and the effects of potentially interfering gases need to be considered.

The materials used in this work are electro-deposited films of tetrabutylammonium bis(1,3-dithiol-2-thione-4,5-dithiolate) nickelate, $Bu_4NNi(dmit)_2$. Metal-dmit complexes have been of interest for a number of years, particularly for their application as organic superconductors[6]. The potential for these materials in a range of devices has been considered, including field effect transistors[7], electroluminescent devices[8] and gas sensors[9]. The latter devices have utilised the fact that these complexes can exist in at least three different oxidation states, namely, -2, -1, and 0 as well as in a non-integral, mixed valence form, $Ni(dmit)_2^{0.29-}$. The room temperature conductivities of $Ni(dmit)_2^0$, $Bu_4NNi(dmit)_2$ and $(Bu_4N)_{0.29}Ni(dmit)_2^{0.29-}$ are 3.5×10^{-3}, 3×10^{-8} and 10 Scm^{-1} respectively[10]. As a result, the redox behaviour is crucial to understanding the interaction between the complex and an ambient gas. In addition to the redox behaviour of these materials, the structure and morphology of the films are particularly important in gas sensing applications and this paper examines the effects of the deposition parameters on the morphology of thin films of tetrabutylammonium bis(1,3-dithiol-2-thione-4,5-dithiolate) nickelate [$Bu_4NNi(dmit)_2$]. The structure of $Bu_4NNi(dmit)_2$ showing the different valencies is given in Figure 1.

2 EXPERIMENTAL DETAILS

In order to enable the resistance of the films to be monitored in the presence of various gases, a gold interdigitated (ID) electrode was fabricated. A thin layer of nichrome (0.3 mm) was RF sputtered onto a clean glass microscope slide and this was followed by a layer of gold (approximately 0.2 µm thick). The required electrode configuration was fabricated using photolithographic techniques and the gold/nichrome was etched using a potassium iodide-iodine mixture. A more detailed description of the device fabrication process is given elsewhere[11].

The electro-depositions were carried out either under potentiostatic conditions using a Thompson Ministat Potentiostat or under potentiodynamic conditions using an EG&G Model 273 Galvanostat. All depositions were carried out in a two compartment, 40 cm³ electrochemical cell, using a platinum wire counter electrode (cathode); a saturated calomel electrode was used as the reference electrode and all potentials in the text are quoted against this reference. The working electrode was either an ID gold grid, a 4 cm² platinum flag or a fluorine-doped tin oxide coated glass slide with a room temperature conductivity of 4.5×10^2 Scm^{-1}. When

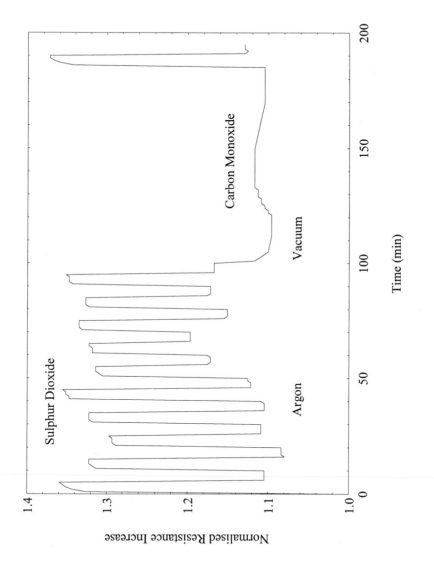

Figure 2 *Response of device in SO₂ and CO*

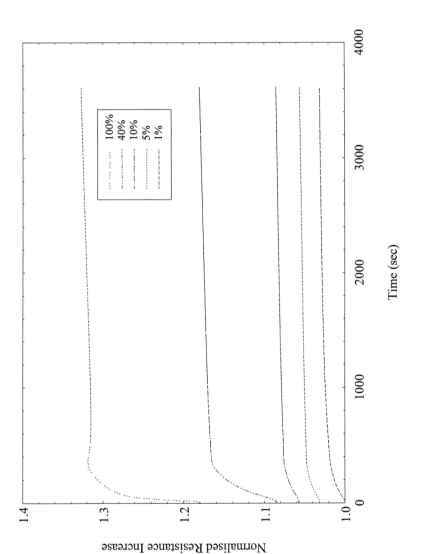

Figure 3 *Resistance increase of electrodeposited film on exposure to varying SO$_2$ concentrations*

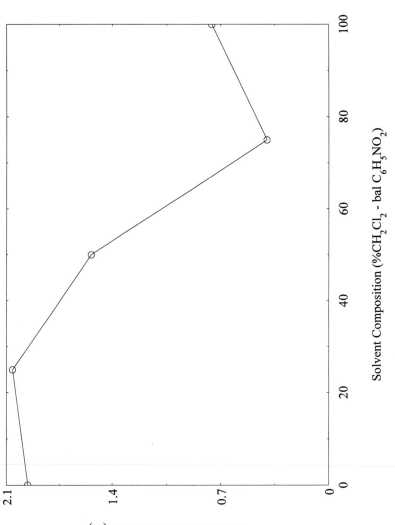

Figure 4 *Response of devices prepared in various mixtures of nitrobenzene and dichloromethane*

(a)   (b)

(c)   (d)

Figure 5 *SEMs of electrodeposited Bu$_4$NNi(dmit)$_2$ on SnO$_2$ coated glass in solvent mixtures: (a) nitrobenzene, (b) 75/25 nitrobenzene:dichloromethane, (c) 75/25 nitrobenzene:dichloromethane, and (d) dichloromethane*

the ID electrodes were used, the two contact pads were connected with Electrodag 915 to short circuit the electrode making both halves of the system anodic. Prior to use, the tin oxide coated glass electrodes were cleaned with Decon 90 in an ultrasonic cleaner, rinsed in acetone and propan-2-ol and blown dry in nitrogen. The platinum electrodes were rinsed in aqua-regia and then cycled for 3.5 minutes in 1M H$_2$SO$_4$ between -1.0 and 1.0 V (vs SCE). The Au ID electrodes were cycled in 10^{-2}M H$_2$SO$_4$ for 1.5 minutes. The treatment in acid was to activate the electrode surface and improve the adhesion of the film. Films were prepared from solutions of 10^{-2}M Bu$_4$NNi(dmit) in C$_6$H$_5$NO$_2$, CH$_2$Cl$_2$ or CH$_3$CN and were carried out at 20°C for 30 minutes. A charge of approximately 450 mC was passed in the deposition process. After the films were deposited, they were rinsed in n-hexane (to remove any nitrobenzene) and vacuum dried at 70°C for 8 hours.

(a)   (b)

(c)   (d)

Figure 6 *SEMs of electrodeposited Bu$_4$NNi(dmit)$_2$ on Au ID electrode in solvent mixtures: (a) nitrobenzene, (b) 75/25 nitrobenzene:dichloromethane, (c) 75/25 nitrobenzene:dichloromethane, and (d) dichloromethane.*

Scanning electron microscopy was carried out using an ISI DS130S SEM. Conductivity measurements were made with a four point probe technique at room temperature. Solutions containing 10^{-3}M Bu$_4$NNi(dmit)$_2$, 10^{-2}M Bu$_4$NPF$_6$ in C$_6$H$_5$NO$_2$, CH$_2$Cl$_2$ or CH$_3$CN were used for the cyclic voltammetry. All solutions were degassed with nitrogen for 20 minutes prior to use.

3 RESULTS AND DISCUSSION

3.1 Gas Sensitivity

Films from Bu$_4$NNi(dmit)$_2$ were deposited onto the Au ID electrodes, both in the presence

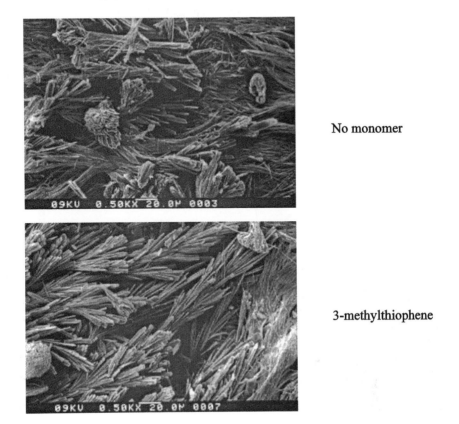

No monomer

3-methylthiophene

Figure 7 *SEMs of electrodeposited $Bu_4NNi(dmit)_2$ with and without 3-methylthiophene on Au ID*

and absence of 3-methylthiophene, and subsequently exposed to sulphur dioxide. The changes in resistance of the films were recorded and the responses of both films were identical. The response of the film prepared without 3-methylthiophene to pulses of SO_2 followed by pulses of argon is shown in Figure 2. It can be seen that the resitance increased by a factor of 1.35 on admission of SO_2, and returned almost to its base level on flushing with argon. After 10 cycles, the chamber was evacuated to 5×10^{-3} torr for 20 minutes when carbon monoxide was introduced. On admission of CO, the resistance of the film increased slightly and this was probably due to pressure effects rather than an electronic interaction. Finally, SO_2 was admitted in to the chamber and the device response was unaffected by the presence of CO.

Identically prepared devices were then prepared and exposed to various concentrations SO_2 in argon. From Figure 3, the response to 1, 5, 10, 40 and 100% SO_2 can be seen: the response time (defined as the time taken to reach 90% of the final value of the resistance) was in the region of 180 seconds.

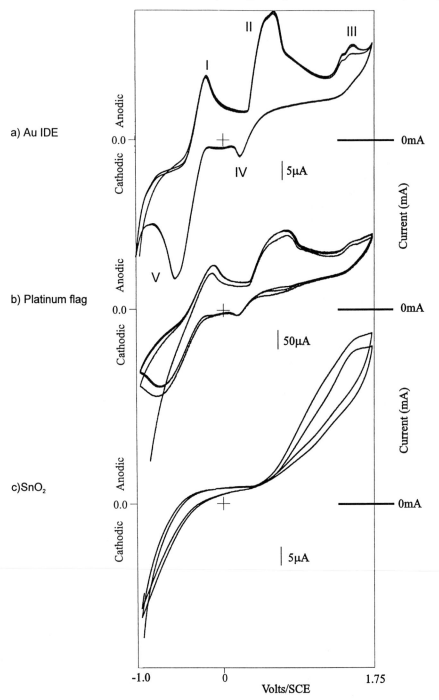

Figure 8 *Cyclic voltammograms of Bu₄NNi(dmit)₂ in nitrobenzene at different electrodes: a) Au ID, b) Pt, c) SnO₂*

Devices were also prepared from CH_2Cl_2 and $C_6H_5NO_2$ and a range of mixtrues of the two solvents and exposed to 1% sulphur dioxide (balance argon). From Figure 4 it can be seen that the response of films prepared in nitrobenzene is a factor of four higher than those prepared in dichloromethane. These devices also showed an increase of 0.5% in 1000 ppm SO_2.

3.2 Scanning Electron Microscopy

Scanning electron micrographs of films deposited at 1.7 volts onto SnO_2 coated glass, in $C_6H_5NO_2$, CH_2Cl_2, 50/50 and 75/25 mixtures of $C_6H_5NO_2$ and CH_2Cl_2 are shown in Figure 5. The film prepared in CH_2Cl_2 consisted of thin, regular platelets whereas in the $C_6H_5NO_2$-CH_2Cl_2 mixtures the film still consisted of the platelet structures but they were not as regular in appearance as those prepared in pure CH_2Cl_2. Films prepared in $C_6H_5NO_2$ had a completely different morphology, consisting entirely of thin needle-like crystals of 300 μm in length. The surface area of the films deposited in $C_6H_5NO_2$ would be much higher than for the films deposited in CH_2Cl_2, which is an important consideration in gas sensor applications and demonstrated by the improved response of devices prepared in nitrobenzene as seen in Figure 4.

Films prepared under identical conditions onto the Au ID electrode were then examined under the electron microscope and are shown in Figure 6. As with the films prepared onto the SnO_2 substrate, the deposition from $C_6H_5NO_2$ consisted of the thin needle like structures whereas the film prepared from CH_2Cl_2 still showed the platelet structure seen on the SnO_2 electrode, although the platelets were much more like those seen on SnO_2 from the 50/50 mixture. Films deposited from the $C_6H_5NO_2/CH_2Cl_2$ mixtures onto the Au ID electrode showed the needle morphology seen in films deposited from $C_6H_5NO_2$ onto the SnO_2. The most likely explanation is that the much higher conductivty of the Au ID (4.5×10^5 Scm^{-1} cf 4.2×10^2 Scm^{-1} for SnO_2) is influencing the film morphology.

From Figure 7 it can be seen that the presence of 3-methylthiophene in the deposition solution had little influence on film morphology at the Au ID electrode. As the electron transfer process would be slower at the SnO_2 electrode than at the Au ID electrode, due to the lower conductivity, this may have allowed the 3-methylthiophene to interfere with the electron transfer process during the Ni(dmit)$_2^-$ oxidation and thus affect the film morphology.

3.3 Cyclic Voltammetry

Cyclic voltammograms of Bu$_4$NNi(dmit)$_2$ in $C_6H_5NO_2$ are shown in Figure 8 at Au ID, Pt and SnO_2 electrodes. The potential sweep was started at -1.0 volts and ramped to $+1.75$ volts at a sweep rate of 25 mVs^{-1}. The voltammograms were very similar for both the Au ID and the Pt flag electrode with a number of distinct peaks due to the oxidation and reduction reactions of the Ni(dmit)$_2^-$ species. In these experiments the starting material was Bu$_4$NNi(dmit)$_2$, i.e. the Ni(dmit)$_2^-$ ion, so at the starting potential of -1.0 volts the Ni(dmit)$_2^-$ adjacent to the electrode surface was immediately reduced to the dianion, Ni(dmit)$_2^{2-}$. As the potential was made more positive, the Ni(dmit)$_2^{2-}$ was re-oxidised to the Ni(dmit)$_2^-$ and this is illustrated by peak I at -0.17 volts. A sharp increase in the current was seen when the potential reached 0.29 volts and this is indicative of the deposition of a conducting species onto the electrode surface. At 0.62 volts, a second peak, peak II, was observed which was ascribed to the oxidation of Ni(dmit)$_2^-$ to the mixed valence[6], Ni(dmit)$_2^{0.29-}$, species. Peak III at 1.53 volts was due to the oxidation of the $Ni(dmit)_2^{0.29-}$ to the neutral $Ni(dmit)_2$ species. The scan was reversed at 1.75 volts and no peak corresponding to peak III was observed. This means that the oxidation of the $Ni(dmit)_2^{0.29-}$ to the $Ni(dmit)_2$ was electrochemically irreversible.

At 0.20 volts a small peak, IV, was observed and was probably due to the reduction of any remaining $Ni(dmit)_2^{0.29-}$ to $Ni(dmit)_2^-$. It is possible that not all the mixed valence species was oxidised to the $Ni(dmit)_2$ on the forward scan because of the low conductivity of the neutral species. This means that after a layer of the $Ni(dmit)_2^0$ had formed on the electrode surface, a significant overvoltage would be required to oxidise all the $Ni(dmit)_2^{0.29-}$ to $Ni(dmit)_2$. In order to test this hypothesis, the conductivity of the film was measured by removing it from the electrode and pressing it into a disk. The relatively high conductivity (4.2×10^{-2} Scm^{-1}) of films deposited at 1.7 volts supported the suggestion that the film contained a proportion of the high conductivity $Ni(dmit)_2^{0.29-}$. The Lichternecker model of mixtures [12] implies that for a film with this conductivity up to 30% of its composition would be the mixed valence species, in spite of the fact that a deposition at this potential would be expected to produce essentially $Ni(dmit)_2$. Further support is from the C,H,N,S analysis in Table 1 which suggests the presence of the mixed valence species in the films. As the potential scan was swept to more negative potentials peak V was seen at -0.63 volts and this was attributed to the reduction of $Ni(dmit)_2^-$ to $Ni(dmit)_2^{2-}$.

4 CONCLUSIONS

It has been demonstrated that the morphology of electrodeposited $Bu_4NNi(dmit)_2$ can be modified by the use of different solvents, working electrodes and by the addition of various organic additives to the deposition solution and that these different morphologies can significantly affect the performance of devices fabricated using $Bu_4NNi(dmit)_2$. The effects of the solvent composition on the change of resistance of the films on exposure to sulphur dioxide was demonstrated.

Acknoweldgement

The authors would like to thank British Gas and Cookson Technology Centre for supporting this work.

References

1. M.J. Madou and S.J. Morrison 'Chemical Sensing with Solid State Devices', Academic Press Inc., 1989.
2. T. C. Pearce, J.W. Gardner, S. Friel, P.N. Bartlett and N. Blair, *The Analyst*, 1993, **118**, 371.
3. J. Kaufhod and K. Hauffe, *Ber. Bunsenges. Phys. Chem.*, 1965, **69**, 168.
4. J. R. Bates, PhD Thesis, University of Northumbria, 1993.
5. J. W. Grate, S. Rose-Pehrsson and W. R. Barger, *Langmuir*, 1988, **4**, 1293.
6. P. Cassoux, L. Valade, H. Kobayashi, R. A. Clark and A.E. Underhill, *Co-ord. Chem. Rev.*, 1991, **110**, 115.
7. C. Pearson, A. J. Moore, J. E. Gibson, M. R. Bryce and M. C. Petty, *Thin solid Films*, 1994, **224**, 932.

8. G. Williams, A. J. Moore, M. R. Bryce and M. C. Petty, *Thin Solid Films*, 1994, **244**, 936.

9. J. R. Bates, R. W. Miles and P. Kathirgamanathan, *Synthetic Metals*, 1996, **76**, 313.

10. M. Bousseau, L. Valade, J. P. Legros, P. Cassoux, M. Garbauskas and L. V. Interrante, *J. Am. Chem. Soc.*, 1986, **108**, 1908.

11. J. R. Bates, P. Kathirgamanathan and R. W. Miles, *Electronics Letters*, 1995, **31**, 1225.

12. K. C. Pitman, M. W. Lindley, D. Simkin and J. F. Cooper, *IEE Proceedings F*, 1991, **138**, 223.

Contributor Index

This is a combined index for all three volumes. The volume number is given in roman numerals, followed by the page number.

Subject Index

This is a combined index for all three volumes. The volume number is given in roman numerals, followed by the page number.